Digital Water Hydrology
Theory and Practice

Digital Water Hydrology
Theory and Practice

Brajendra Singh

RANDOM PUBLICATIONS
NEW DELHI (INDIA)

Digital Water Hydrology: Theory and Practice

ISBN 978-93-51114-18-5

Published in 2014 in India by

RANDOM PUBLICATIONS

4376-A/4B, Gali Murari Lal, Ansari Road
New Delhi-110 002
Phone : +9111-43580356, 011-23289044, 011-43142548
e-mail: sales@randompublications.com,
info@randompublications.com, randomexports@gmail.com

Reprinted 2019

Type Setting by: Friends Media, Delhi-110089
Digitally Printed at: Replika Press Pvt. Ltd.

Preface

Recent developments in computational models to monitor and predict hydrology, erosion, and landscape evolution by Earth and water scientists rely heavily on the integrity of the digital elevation models (DEMs) available suggested that Earth and water scientists are not fully aware of the limitations of DEMs as sources of spatial information. This concern is not unwarranted, with few published reports on the accuracy of DEM-derived elevation and slope or the effects introduced on geomorphic parameters; DEMs are created, distributed, and used very frequently without any reference to the magnitude of the error implied or to the methods applied to its disclosure or correction. A DEM consists of either a two-dimensional array of numbers that represents the spatial distribution of elevations on a regular grid; a set of x, y, and z coordinates for an irregular network of points; or contour strings stored in the form of x, y coordinate pairs along each contour line of specified elevation. DEMs on a regular grid are the most widely used data structures because of their computational efficiency and lower storage requirements. Furthermore, grid DEMs are used to calculate all other types of digital terrain models whose accuracy is limited by that of the original source data notes that grid DEMs have several disadvantages: (1) they cannot easily handle discontinuities in elevation; (2) the resolution of the mesh effects the results and computational efficiency; (3) grid spacing needs to be based on the roughest terrain in the catchment, resulting in redundancy in smoother areas; and (4) the computed flow paths tend to zigzag, not following the drainage lines, and are hence systematically too long. Generally, raw elevation data in the form of stereophotographs or field surveys and the equipment necessary to process these data are not readily available to potential users of a DEM.

Most users are therefore forced to rely on the DEMs published by government agencies. The most common form of DEM available in Australia, Organisation Europe´enne d'Estudes Photogramme triques Experimentales member countries, and the United States are those produced by digitizing the contours on existing topographical maps , known as "cartometric DEMs." In many countries these contour-based DEMs are interpolated onto a grid. In addition to interpolation errors this process introduces artifacts such as pits or depressions, which interfere with drainage algorithms based on flow in the steepest downslope direction. In the state of New South Wales (NSW), Australia, this pit removal incorporates a drainage enforcement algorithm developed by Hutchinson [1989], which utilizes the blue lines drawn on topographic maps to represent the permanently flowing or major intermittent streams . In recent years, there has been an increasing move toward using automated digital correlation techniques to generate what are known as "photogrammetric DEMs" directly from stereoscopic imagery, especially where contour data are not available or are not accurate enough. These methods create gridded DEMs directly with similar artifacts to the cartometric DEMs, though studies of these errors for hydrologic purposes have not been performed. An additional drawback is that the points which are correlated (used to determine the elevations) are often the tops of trees, buildings, etc., consequently requiring a significant amount of editing by hand to produce a realistic DEM. Few studies have been undertaken on the effects of relative general conclusions.

The book will be an indispensable source for all professionals, researchers and students in this subject and for anyone working in the related areas for acquiring an up-to-date overviews.

I thank all members of my team who have helped in the preparation of the book. My special thanks go to "Random Publication" who have published the book.

—Brajendra Singh

Contents

1

Introduction

"All the rivers run into the sea; yet the sea is not full; unto the place from whence the rivers come, thither they return again." Here, most succinctly, the Biblical writer has expressed what water scientists call the "hydrologic cycle." The details of how this cycle came about are not the concern of the verse writer quoted; for him it was enough to suggest the eternal circle of water. Today those men concerned with the world's water supply must look into the exact details of the hydrologic cycle. Only when we know the nature of our resources can we intelligently plan how to provide water in plenty. That great fictional detective Sherlock Holmes once remarked that a logician could infer Niagara Falls from a single drop of water. Unfortunately, hydrology is far more complex, and logicians are far less astute, than Holmes would have us believe.

The first water was most probably not from the sea, but in the clouds. For simplicity, however, let us begin with the sea, that vast reservoir holding all but a tiny fraction of our water. For man, to whom the withholding of a mere two quarts of the liquid a day can mean death, the quantity of water in the oceans is all but incomprehensible. To number the quarts in the sea would result in astronomical figures. Even the number of *cubic miles* contained is fantastic to those of us not used to dealing in such figures. There are some 327,672,000 cubic miles of water in the oceans. Expressed in a way that is easier to visualise, there is sufficient water in the world that all the dry land could be covered with a column of water five miles high. Of this total, 98.33 percent is sea water. Herein lies the big problem of man dependent on fresh water to drink.

Leonardo da Vinci was a pioneer in pinpointing the true nature of the hydrologic cycle. Among his many careers was that of commissioner of canals for the Milan area, and he learned enough about water to write, "we may conclude that the water goes from the rivers to the sea and from the sea to the rivers, thus constantly circulating and returning, and that all the sea and rivers have passed through the mouth of the Nile an infinite number of times." Egyptians would have been inclined to agree with this, adding perhaps that it seemed that *all* the water passed through the Nile several times during flood season. Da Vinci goes on: "The conclusion is that the saltness of the sea must proceed from the many springs of water which, as they penetrate the earth, find mines of salt, and these they dissolve in part and carry with them to the ocean and the other seas, whence the clouds, the begetters of rivers, never carry it up."

Da Vinci erred very slightly at this point, for there is some salt carried into the atmosphere. In fact, some scientists believe that salt forms nuclei for raindrops, as do ice crystals and dust particles. There is salt in rainwater, but since it is there in a concentration of only about a few parts per million, Da Vinci was essentially correct. The clouds, indeed, are the begetters of rivers. Each year about 80,000 cubic miles of water evaporates from the oceans of the world. Another 15,000 cubic miles evaporates from water on the continents. In due course, this huge amount of water must fall as some sort of precipitation. Because they represent only about one-quarter of the surface of the earth, the lands receive "only" about 24,000 cubic miles of precipitation a year.

We can thank the sun for operating this vast global distillery of fresh water for us. Aiding in circulating the vapour content of the atmosphere are the winds generated by the rotation of earth and by the uneven heating of land and water surfaces. The result is a great mixing that guarantees world tours for individual water molecules.

Sunspots, too, are thought to be connected with rainfall; the hydrologic cycle is not a simple one at all. Some idea of the tremendous power furnished by the sun lies in the fact that to provide an inch of fresh water artificially over one square mile would require the burning of 6400 tons of coal, or its equivalent, in a distillation plant.

Except in rare cases, not more than 10 percent of the moisture content of clouds is precipitated, even though it often seems that the "bottom has fallen out of the clouds." Our share of precipitation in the United States, based on the average annual rainfall of 30 inches, is about 1,564,800 billion gallons a year. This sounds sufficient for even the most wasteful nation imaginable, but much of this rainfall is lost without doing us any good. Within the hydrologic cycle is a smaller internal cycle called "evapotranspiration." Evaporation is a process we are familiar with; transpiration is the process by which plants "breathe out" moisture to the atmosphere.

In humid, cool areas of the country, evapotranspiration gives back to the air about 40 percent of moisture. As we might guess, the percentage goes up in drier, warmer areas, until in the arid Southwest as much as the full 100 percent of precipitation is lost to the atmosphere. On a nationwide average, about 72 percent of precipitation is given back to the air. This means that only 28 percent is retained in or on the land for use by plants and animals. It is wrong, of course, to think that transpiration is a complete waste of water, for without it plants could not exist. However, the result is the loss of water for further use.

After evapotranspiration losses, we in the United States have about 1100 billion gallons available as "runoff" each day. This of course represents a maximum amount, and because people and plants are not always where water is, we cannot hope to be 100 percent efficient in using this "available" water. At present we are using about 320 billion gallons a day. Twice that amount would begin to strain the water budget.

The concept of runoff introduces another small cycle, or "loop," within our overall hydrologic cycle. Part of the water available after evapotranspiration losses becomes "ground water," literally water that is in the ground. This ground water takes between 33 percent and 40 percent of total runoff, and provides a natural reservoir of water for man's use if he can get to it with wells and pumps.

The ancients knew of ground water, and made use of it too, although they were uncertain as to how it came to be where it was. Ground water is not to be confused with the moisture in the soil near the surface; it is deeper down, and stored in aquifers, which are strata of permeable, or porous, rock. Hydrologists point out that underground water does not flow in "rivers" of great size, although this idea still persists in the minds of many people. Obeying natural laws, including the law of gravity, ground water seeks a lower level. Much of it comes again to the surface as springs, and it is estimated that about 40 percent of our rivers are fed in this manner.

Aquifers may be several hundred feet in thickness, and extend for many miles beneath the surface of the earth. Despite the term "water *table*," underground water is not necessarily level as is a body of surface water. For this reason, there are *artesian* wells from which water flows with no pumping. Named for the French province of Artois (Artesium in Roman times), where the first such wells were drilled in 1126, artesian water is under pressure from water at a higher level and will flow until the level drops sufficiently to reduce that pressure. Where this pressure, or "head," of water does not exist, man must lift the ground water in his well with buckets, or pump it up with a mechanical pump or one that is electrically driven.

Estimates of total ground water range from enough to cover the earth's land surfaces to a depth of 100 feet, to as much as several times that. The amount of ground water in the United States is believed to equal the amount contained in the Great Lakes. The beauty of this underground storage system is that it loses little water to evaporation, whereas the Great Lakes lose many billions of gallons each year in this way. The evaporation from the surface of Lake Mead has been measured as some 800,000 acre-feet a year. To gauge how much this is, a single acre-foot of water would provide a family of five with 180 gallons of water apiece each day for a year. The loss from reservoirs has been shown to run as high as 90 percent of the amount taken usefully from the lake.

The concept of water as a mineral, an oxide of hydrogen, is interesting with reference to the use of ground water. If a user draws from his well faster than the ground water can be replenished by precipitation of surface water, the water table drops. This is called the "mining" of water, since it depletes the supply available. Such a drop in the water table is occurring in many places in the United States and the world, although paradoxically a problem in other areas is the rising water table. This strange fact we shall discuss later; for now, let us consider only the maintained or declining amount of water in our aquifers.

Some authorities believe that ground water equals *all* our surface reservoirs and lakes, and not just the Great Lakes. The water stored in the ground is estimated to equal ten years of precipitation, or thirty-five years of runoff. This is a comforting thought, along with the knowledge that we are using only about one gallon of ground water for each four gallons of surface water for domestic, agricultural, and industrial use.

Generally, the water from wells or springs is purer than that in our lakes and rivers. It has been protected from contamination by natural causes and also from man-made pollution (with exceptions, of course). Besides this, it has been filtered through rock, sand, or other material that removes much of the contaminating matter.

An accurate description of ground water is the part of subterranean water that occurs where all the pores in the containing rock materials are saturated; that is, they hold all the water it is possible for them to hold. This zone of saturation in some cases extends all the way to the surface of the earth, as, for example, in seep areas, marshy ground, lakes, and so on. In other locations there is a layer, called the "zone of aeration," above the saturated aquifer, ranging from inches to hundreds of feet. There is water in this layer, some of it called capillary water, since it is held in the soil against the pull of gravity. But wells cannot make use of this relatively small amount of water, and Vitruvius of old was not correct in stating that moist surface soil always indicated true ground water. Moisture in the soil, while not nearly the quantity of true ground water, has still been estimated at enough to cover all the land surface of the earth to a depth of about 4 inches.

Through scientific tracing techniques, the course of underground water can be followed. Using dyes, experts have found wells being polluted from nearby sources. They have also found that water can move more than 300 miles through an aquifer. Ancient well diggers managed to dig through earth, and sometimes solid rock, to depths greater than 1000 feet. Such feats required many generations to accomplish, as we have noted. Today modern well-digging equipment can sink wells deeper than a mile, and up to five feet in diameter, in a relatively short time. When it is possible, a well is located on a high point of ground so that it can feed a large area by gravity flow, without need of an elaborate pumping system.

It is generally accepted by scientists today that ground water is replenished by the hydrologic cycle we have described; that is, by water precipitated from the atmosphere. There are, however, other theories that are interesting to consider, even though they are not widely accepted. One theory is that of subterranean condensation of water vapour from air that enters the earth. It is well known that water condenses on a cool surface, and perhaps to some extent moist air blowing on the ground adds to its water content.

Another theory holds that some ground water is "juvenile," or new, water that is constantly being manufactured from the hydrogen and oxygen that form part of the rocks far beneath the surface. Great pressures do exist at such depths, and perhaps water is formed by processes we can only guess at. Such new water is sometimes called magmatic water or volcanic water, as it is believed to be associated with magma, or molten rock from deep in the earth. Hot springs are also thought by some authorities to be at least partly juvenile or volcanic water. There is another, less controversial type known as "connate" water. This is water, much of it salty, that was trapped ages ago in folds of rock as the earth heaved and twisted during its formation. Such deposits of connate water are occasionally released by earthquakes, and add to the total water supply.

The hydrologist Oscar E. Meinzer has said, "The supply of water is not a fixed and immutable quantity, but is decreased by hydration and increased by dehydration of the rocks." Man himself adds to and detracts from the total water supply in such process as the combination of hydrogen and oxygen (in the welding torch, for example) in concrete work wherein some water is relieved of its hydrogen content and so on. But these are tiny amounts when compared with the water nature itself rearranges each year. For example, it has been estimated that plants annually reduce some 535 billion tons of carbon dioxide from water. One writer goes further, pointing out that this means the destruction of about 600 billion tons of water each year, a loss that would deplete the oceans themselves in about 21/3 million years! Since plants have been accomplishing the photosynthesis reaction far longer than 2 1/3 million years, and the seas are still here, he assumes that new water in that amount must be continually added to the "hydrosphere."

Except for traces of organic and inorganic matter, rainwater is pure. Man has added a pollutant in the form of radioactive particles from nuclear-bomb fallout, but precipitation reaching the ground is still very low in foreign matter. However, as it percolates, trickles, or flows back to the sea from whence it came, water washes salt and other minerals from the soil. This salt remains in the sea, adding to its saltiness year after year. However, ground water is endangered by the encroachment of salt water. This happens in many inland states as well as those near the sea, and steps must be taken to prevent further harming of our freshwater supplies.

Snow is part of the hydrologic cycle. This water in solid form is kept in storage on mountain watersheds, and adds to runoff during spring thaws. There are other forms of water that are important, although not so important as rain itself. These include fog, dew, and mist.

Fog is sometimes called low cloud, and if you run into clouds in the mountains you may be tempted to call them high fog. Fog occurs at ground level, and is caused by a number of factors, including humidity, temperature (both of the air and of the ground or water), air pressure, and wind velocity. In some coastal areas fog is a beneficial addition to precipitation. Tomatoes and other crops have been grown in strips of land along the California seacoast, irrigated almost entirely by fog. Whether or not forests make good use of fog is a controversial point. Some authorities claim that they do, accumulating much moisture from the fog, and using it not only for their own benefit but also for that of underbrush beneath them.

Mist is heavier than fog, and amounts to slight precipitation. Dew is something different than either. Dew contributes considerably to agriculture, and it is believed that it can be exploited far more than it has been so far. Ages ago, clever farmers irrigated plants and trees with dew by building rock cairns to condense moisture out of the air. Even on a clear day, the atmosphere contains much water, and if there were a way to extract it the desert might have a supply of water.

Dr. J. H. Dannies, a German engineer has for years experimented with equipment for extracting moisture from the atmosphere. This is done by cooling the air below its dew point. In this way Dannies has produced drinking water from the air in the Sahara Desert. More recently, Israel has investigated the possibilities of using the dew to supplement that country's meager rainfall. Tests show that in a year's time a total of 1.5 inches of water collects on the cold desert surface as dew. However, theory, and tests using methods of increasing condensation with a cold surface, indicates there should be 15 inches of dew available. Added to a normal rainfall of 5 inches, this would be a creditable total of precipitation for Israel. There are some hydrologists who claim that beneath the arid Sahara Desert in Africa is a great stratum of water-bearing rock and that all that is needed is a system of wells to tap it to make the desert once again the lush land it was described to be in ancient literature.

Floods and Drought

Water is important to life. Moreover, it must be available in the proper amounts. From the Genesis flood on, history is filled with tragic records of loss of life when rampaging water went out of control and drowned luckless human beings and animals. In 1223, for instance, Friesland in Northern Europe was flooded and 100,000 lives were lost. In 1521, Holland lost 100,000 people by drowning when the sea rose over the land. A worse flood came in 1642 when China's Hwang Ho went out of bounds and killed 300,000 people. Four years later, another 110,000 died in a flood that ravaged Holland and Friesland. In 1876, a tidal wave killed an estimated 200,000 at Bengal, India. China was struck again in 1887, and hundreds of thousands of people died when the Yellow River (Hwang Ho) flooded. Another 100,000 Chinese drowned in 1911 when the Yangtse left its banks. Another disaster involving a dam failure came as recently as 1963 when one in the Italian Alps killed 1800 people.

Figure : *Dry Earth in the Sonoran Desert*

Such numbers of casualties make floods in our own country seem less terrible by comparison, but Americans have had their share of disaster. In 1889, a dam burst in Johnstown, Pennsylvania and over 2200 people lost their lives. Another 450 died when a dam collapsed near Los Angeles in 1928. From time to time, our country still suffers from disastrous floods. In 1951, the lower Kansas River went out of control, affecting some half a million victims. On the average, about

75,000 people are forced from their homes each year by rampaging floodwaters. Surprisingly, of the last three decades of floods that cost the United States billions of dollars plus many lives lost, the worst floods came in 1936 to 1938 when the Midwest drought was at its height. Just as too much water can be catastrophic, so can too little. One of the first recorded famines was that in Rome in 436 B.C. Ironically, thousands of the starving committed suicide by drowning themselves in the sea. Many countries were struck by the famines that occurred during the Middle Ages, in 879 and 1162. Millions have died in India, a land constantly blighted by drought, and in 1877 there were about ten million Chinese who died of starvation.

The causes of drought are not known with any certainty. It is believed that weather patterns are cyclical and that perhaps they are affected by sunspots. However, there seems no regularity in the pattern of drought. In the terrible "dust bowl" days of the 1930's, rainfall in the United States was down only about two inches from the normal average of 30 inches. Perhaps man himself is a contributor to drought and famine.

At its prime, the civilization of ancient Mesopotamia provided a good life for some 40 million people. After its fall, at the hands of invaders, less than five million inhabitants could survive. This was most likely due to the fact that the invaders destroyed the already overworked irrigation system that had made Babylon the mighty city of the ancient world. Canals filled with silt, and improperly drained fields became clogged with minerals so that they would no longer grow crops. On the North American continent, Indians of the Southwest had a flourishing civilization built on water culture. But the Babylonian pattern was repeated, and the Pimas abandoned the lands they had laboriously watered with canals. In modern times, Arizona has repeated the cycle, adding to the salinity of the well-named Salt River Valley until it is again a problem area. In the dust bowl of the Midwest, overgrazing and erosion permitted winds to carry away precious topsoil. Not just lack of water, but the plows too ruined the land.

Following World War II, parts of the United States experienced a drought that lasted a decade. The Northeast was hard hit, and the water shortage that plagued New York in 1949 focused interest on the water problem. Drought was not new, but combined with ever-increasing demand for water it created a new situation to be reckoned with. Late in 1961, another drought began in the Northeast. By 1965 it was recognised by the Water Resources Council in Washington,

D.C., as "the most intense drought in the history of that area." In a report to the President late in July, 1965, the council noted a condition of "extreme drought" in the northern half of New Jersey, the northeast quarter of Pennsylvania, the southeast half of New York, all of Vermont, all of New Hampshire, parts of Maine, and the western half of Massachusetts.

Some 300,000 square miles of area were affected by the drought. Included were heavily industrial and urban areas from Philadelphia to New York. Besides restricting the amount of water available for domestic and industrial use, the drought increased forest-fire hazards, hurt crops, and threatened further encroachment of salt water on freshwater supplies. Most alarming of all, it pointed out that there is a water crisis *now,* and not merely the danger of one in the future.

Today's hydrologists are not so miraculously skilled as Sir Conan Doyle suggested. From a single drop of water they cannot infer Niagara Falls or even a smaller part of the hydrologic cycle. They are, however, much further along the road to that doubtful goal than were scientists and engineers of a few centuries ago. Man now knows the workings of nature's "waterwheel," at least in broad outline. Instead of the erroneous theories of the Greeks and the animalistic view of hydrology that Descartes advanced, water experts today possess a reasonably accurate blueprint of water in motion, as well as many bits of data both qualitative and quantitative. And this is well, of course, because as our knowledge of water has grown, so too has the water problem.

Controlling and Restoring the Waters

Water is a source of life, power, comfort, and delight, a universal symbol of purification and renewal. Like a primordial magnet, water pulls at a primitive and deeply rooted part of human nature. More than any other single element besides trees and gardens, water has the greatest potential to forge an emotional link between man and nature in the city. Water is an element of wondrous qualities. It is a liquid, a gas, or a solid. It absorbs energy and transforms it. It transports other elements in suspension and solution, shaping the landscape and nurturing life. It permeates the terrestrial environment — air, earth, and all living organisms. Pure, in the right place, and at the right time, water is an essential resource; impure, and at the wrong place and time, water is a life-threatening hazard.

An abundance of potable water is a crucial concern of all cities. To this concern, we owe some of the greatest architectural monuments of human history and some of the most impressive engineering works: the aqueducts of Rome and Nîmes and the qanÜ of Persia. Eleven aqueducts, bringing water from ten to fifty-nine miles away, supplied Imperial Rome with approximately 35 million gallons of water per day. The aqueducts delivered water to reservoirs from which it was distributed to all parts of the city. Pliny described this feat as one of the greatest achievements of Roman civilization: "But if anyone will note the abundance of water skillfully brought into the city, for public uses, for baths, for public basins, for houses, runnels, suburban gardens and villas; if you will note the high aqueducts required for maintaining proper elevation; the mountains which had to be pierced for the same reason, and the valleys it was necessary to fill up; you will conclude that the whole terrestrial orb offers nothing more marvellous."

Water availability not only determined the site of ancient cities, but also the arrangement of buildings within them. More than 3,000 years ago, the Persians first built qanâts — tunnels many miles long and up to three hundred feet deep — to carry water from mountain slopes to cities at the desert's edge. The hydraulic gradient was a measure of status. The houses and fields of the wealthy were uphill and received the water first. They used the water and passed it on. The poor, whose homes and fields were at the lowest elevations, received the water last. Stone-lined conduits, similar in design to their ancient predecessors, provide many Iranian towns with water today. The wealthy residential districts are still elevated, the poor districts depressed.

Aristotle recognised that an ample water supply was essential to both military security and health: "There should be a natural abundance of springs and fountains in the town, or, if there is a deficiency of them, great reservoirs may be established for the collection of rain water, such as will not fail when the inhabitants are cut off from the country by a war for the elements which we use most and oftenest for support of the body contribute most to health, and among those are water and air."

Urban civilizations have long grappled with the problems of water supply and use, sewage disposal, storm drainage, and flood prevention. Together, these have probably received more sustained attention throughout history than any other single urban problem. There is no lack of models for successful resolution to these problems. Urban cultures that arose in the arid and semiarid climates of Persia and the Mediterranean have developed a landscape art that both conserves and displays water. Cities like Denver, Colorado, that have reclaimed their rivers for recreation, while implementing flood prevention and water quality measures, illustrate the many social and economic benefits such projects generate. Cities that have exploited the flood storage and water treatment potential of wetlands demonstrate how parks and urban wilds can serve many uses. Most of these models, however, consist of solutions to a single aspect of the water problem: either storm drainage and flood control, sewage treatment, or water supply and conservation. The comprehensive, natural drainage system of Woodlands, Texas, a new town thirty miles north of Houston, exemplifies the advantages of considering storm drainage, flood control, water quality, and water conservation in a single scheme. Whatever the scale from the design of a drain or a fountain to a plan for an entire metropolitan region the key to devising efficient, effective, and economical solutions is an understanding of the many ways water moves through the city.

Water in Motion

"All the rivers run into the sea, yet the sea is never full; unto the place from whence the rivers come thither they return again." The hydrologic cycle is a grand process by which rain falls on the land, is absorbed by the earth and the plants that grow in it and runs into streams and oceans, then evaporates, returning once more to the air. The power of the sun and the force of gravity drive the hydrologic cycle. The way water moves through the hydrologic cycle determines the distribution of water supplies, the occurrence of floods, and the fate of contaminants disposed of to the air, water, or land.

Only a fraction of the rain that falls on rural woods and fields runs rapidly into streams, rivers, and lakes. Leaves intercept some rain, and soil soaks up much of the remainder. Of the water that soaks into the soil, some is sucked up by plants and later returned to the atmosphere via evapo-transpiration, some evaporates directly from the soil's surface, and the remainder moves slowly through the soil as ground water. Ground water may eventually intersect the land's surface at stream beds and springs or may remain deep beneath the surface in vast underground reservoirs or aquifers. Only on steep slopes, on bare rock or ice, or when the soil is saturated, does water run off the ground's surface. The great capacity of soil and the organisms within it to absorb water and to filter and use the elements suspended or dissolved within it prevents floods, protects water quality, and conserves and restores water supplies. Traditional urban storm drainage systems short-circuit this portion of the hydrologic cycle, with disastrous results. Some cities have attempted to re-establish that link in the cycle by retaining stormwater and permitting it to infiltrate the soil; others have merely detained stormwater until the flood hazard has passed and water can be treated or safely released.

Some of the sources of water pollution — factories, sewage treatment plants, erosion from construction sites, urban runoff from storm sewers, and the fallout of dust from the air — can be pinpointed to the discharge from a specific pipe or ditch; others are more diffuse. "Point" sources are readily monitored and regulated. One can identify and measure the specific pollutants discharged, plot the precise location where they enter the water, and, given the depth and size of the water body and the circulation pattern of the water within it, predict the likely pattern of their distribution. New "point" sources, like factories or treatment plants, can be located in areas with adequate water circulation, distant from swimming beaches.

As more and more industries and municipalities conform to federal water standards, "nonpoint" sources, like air pollution and urban runoff, will become more critical water pollution problems. Nonpoint sources are extremely difficult to regulate except by collecting and treating all stormwater. Flood prevention strategies that involve the retention or detention of stormwater promise to benefit water quality, since most of the suspended solids settle out in standing water and many of the nutrients, oil, and grease are filtered out as water moves through the soil.

Storing Floodwaters

The past decade has seen a profusion of outstanding, innovative approaches to flood control by American cities. Rooftops, plazas, parking lots, and parks have been designed to store stormwater, and woods and wetlands in the headwaters preserved for their natural storage capacity, thereby reducing floods and the cost of storm drainage systems and, in some cases, permitting the treatment of urban runoff. This has generally been accomplished with little or no extra construction cost, with minimal inconvenience, and has resulted in the acquisition of new recreation land. The key to preventing floods and minimising the destruction they wreak lies in a dual strategy of storing stormwater until flooding peaks and eliminating obstacles to floodwaters within the floodplain. These principles apply whether designing a rooftop to pond and detain rain water or designating undeveloped urban wetlands as parkland to soak up and hold water in soil and plants; whether designing a pedestrian bridge so as not to block debris in floodwaters or establishing land use and building regulations in floodplains.

Floodwater storage and recreation are compatible in large urban parks. Parks that exploit the natural flood storage capacity of floodplains capture the water's edge for the public landscape. The recent profusion of urban parks that serve multiple purposes of flood control, water quality improvement, and recreation do not reflect a new idea, but rather the rediscovery of old solutions.

Many nineteenth century and early-twentieth-century parks, now valued for their access to urban rivers and lakefronts, were originally designed as flood control and water quality projects. Landscape architects and urban historians regard Boston's "Emerald Necklace" park system as a landmark in American park planning, but few appreciate that a third of the system was designed as a flood control and water quality project and not primarily for recreation.

The designer, Frederick Law Olmsted, created the Fens and the Riverway to combat the flooding and pollution problems of Boston's Back Bay tidal flats; public recreation was an incidental benefit and Olmsted himself objected to the use of the word "park" for the Fens, since he did not consider it an appropriate spot for any recreation beyond a stroll or drive along the border of the marsh. The statement printed on Olmsted's 1881 map, "General Plan for the Sanitary Improvement of the Muddy River," declares this intent:

> The primary design of the scheme here shown is to abate existing nuisances, avoid threatened dangers and provide for the permanent, wholesome and seemly disposition of the drainage of Muddy River Valley. This is proposed to be accomplished chiefly by embanking, contracting and deepening the existing creek and ponds and excluding sewage and tides. The secondary design is to make use of the embankments required for the above purpose to complete the promenade here shown, of which the Common, Public Garden and Commonwealth Avenue would form about one-third already prepared and in use, and the Back Bay, now half-formed, and in progress, another third...

Until recently, historians have admired Olmsted's Boston park system chiefly for its connection of the central city with outlying suburbs, in a series of parks and connecting parkways, forgetting the flood control and water quality purpose that portions of it originally served. Olmsted designed the Fens as an irregularly shaped depression scooped out of the tidal flats. The configuration and size of the thirty-acre basin permitted the amount of water to double without raising the water level more than a few feet; during floods, twenty additional acres could be covered with water.

Gently sloping banks and an irregularly shaped edge reduced waves. A tidal gate at the entry to the Charles River regulated the flow of the tides to prevent flooding and to enhance flushing of the basin. Part of Olmsted's plan was the restoration of the former salt marsh; he planted the banks of the basin with plants that could tolerate both salt and brackish water and withstand changing water levels.

Olmsted felt that the juxtaposition of salt marsh and city would be novel, certainly, in laboured urban grounds, and there may be a momentary question of its dignity and appropriateness... but it is a direct development of the original conditions of the locality in adaptation to the needs of a dense community. So regarded, it will be found to be, in the artistic sense of the word, natural, and possibly to suggest a modest poetic sentiment more grateful to townweary minds than an elaborate and elegant gardenlike work would have yielded.

Portions of the Fens were planted by 1884 and within ten years had the look of a landscape that had always been there. The rapid success was largely due to the sheer quantity and diversity of vegetation planted: more than 100,000 shrubs, vines, and flowers in one area of two-and-a-half acres.

The Muddy River flows into the Fens, its current alignment and shape the nineteenth century's artificial creation. The banks of the Muddy River were regraded, lined with walkways, crossed by bridges for pedestrians and vehicles, and planted with grasses, shrubs, and trees to form the "Riverway".

Like the Fens, within a few decades of construction, the Riverway had the appearance of a natural floodplain penetrating the city. Depressed below street level, with steep, wooded banks between the roadway above and the path below, it is still a retreat in the middle of modern Boston. The Muddy River survives more intact than the Fens. After the Charles River Dam was constructed in the early twentieth century, the salt marsh declined, the Fens lost the aid of the tides in enhancing water circulation, and ultimately became a dumping ground for fill from the subway and other projects.

Figure : *The Muddy River in early Spring*

Chicago, built on a flat plain only slightly higher than Lake Michigan, has been plagued by drainage and flooding problems throughout its history and has responded with ingenious solutions. In the mid 1800s, Chicago raised its street level twelve feet, jacked up and elevated existing buildings, and installed a new storm sewer system. After 12 percent of the city's population died in 1885 from cholera, typhoid, and dysentery contracted from a polluted water supply, Chicago established an autonomous regional agency, the Metropolitan Sanitary District of greater Chicago. For nearly a century, this organisation has coordinated Chicago's flood control, storm drainage, and sewage treatment. Chicago has a combined sanitary and storm sewer system and now uses stormwater detention basins located throughout the city in floodplains to detain stormwater before it reaches storm sewers, along with an extensive system of deep, underground tunnels to store the overflow from the sewer system until it can be treated. The Melvina Ditch Detention Reservoir is one of the many large detention basins operated by the Metropolitan

Sanitary District and used for both flood control and recreation. Steps lead down the basin's side slope to playfields and volleyball and basketball courts in the bottom of the basin. Children ski and toboggan down the slopes of a large earthen mound at the corner of the basin and skate on an ice rink created by flooding a large, paved area near the basin's inlet. When flooded, the reservoir holds 165 acrefeet of water.

Parking lots, which account for much of the open, paved land in American cities, can also be designed to detain or even retain stormwater, as one was at the First National Bank in Boulder, Colorado, where a section of the lot can hold up to two feet of water. Consolidated Freightways in St. Louis, Missouri, constructed its parking lot to detain storm flows and netted a $35,000 savings in the cost of the storm drainage system. Outside the downtown, in less dense parts of the city, it may be preferable to retain water long enough for it to infiltrate the soil. Porous pavement porous asphalt, modular paving, and gravel over well-drained soils or in combination with dry wells will permit more rainfall to soak into the ground rather than run off into storm sewers. A pavement of lattice concrete blocks, with soil and grass in the interstices, is widely used in European cities, and has been employed in parts of some American cities such as Los Angeles and Dayton.

Restoring and Conserving Water

The restoration of water is also an essential function. A sewage treatment facility can be attractive and, in certain phases of its operation, compatible with recreation. In 1967, after the state of Michigan threatened to cite the city of Mt. Clemens for pollution of the Clinton River, the city combined a new sewage treatment system with a park. Combined storm and sanitary sewers comprised 90 percent of the Mt. Clemens sewer system, and sewer overflows during rainstorms had been responsible, in part, for pollution of the river. After several years' study, the city determined that collecting, storing, and treating the combined overflow was more feasible, more efficient, and less costly than separating the storm and sanitary sewer systems, and that it also offered an opportunity to create new parkland. The city constructed its new sewer overflow treatment facility with three small lakes and a park on a former sanitary landfill site. Sewer overflows remain in the first lake for anywhere from one to four days, until they can be treated in the processing building, then the water is released for aeration to the second lake for an additional seven days.

By the time the treated effluent reaches the third lake, 2.3 acres and 9 feet deep, it is appropriate for boating and fishing and for irrigating the park's landscape. In winter, when the third lake freezes, it is used for skating and ice hockey. The city plans to stock it with fish and construct a dock for summer recreation.

***Figure :** Drip irrigation system in New Mexico*

Arcata, California, exploits the properties of plants and soil to assimilate wastes, by using a wetland as part of its wastewater treatment process. Arcata renovated and reconstructed a degraded, existing wetland adjacent to its sewage treatment plant to enhance the quality of its water after treatment. The reconstructed wetland serves other functions including wildlife habitat and recreation. Other cities, including Austin, Texas, have experimented with natural or constructed wetlands to treat sewage effluents. Because wetland or aquatic plant systems to treat wastewater require more land area than conventional treatment methods, they are likely to be most appropriate for small to moderate-sized cities. The danger of introducing concentrated heavy metals and toxicants into the food chain rules out the use of such systems when effluent is heavily contaminated by these pollutants. Wetland treatment systems will be most useful in providing advanced treatment where traditional chemical methods are too costly, and they are likely to become more common as current successes become better known.

Sewage treatment can both conserve water and create an aesthetic resource. Five miles out of Santa Fe, New Mexico, a resort named The Bishop's Lodge has built a package sewage treatment facility to provide irrigation water for the resort's pasture and garden. It forms an unusual amenity in this water-poor landscape. Treated wastewater tumbles down waterfalls and cascades through sculpted channels and streams from high ground into a large pool. These "seven magic pools" provide tertiary treatment to the wastewater by aerating it and exposing it to sunlight. The water cascades a hundred feet to the resort's entrance; landscaping and earth mounds screen the treatment plant from view. Water conservation is an important benefit. Formerly, The Bishop's Lodge used 10,000 gallons of well water per day to irrigate its lawns, nearly one-third of the total daily consumption. Irrigation water now consists entirely of treated sewage effluent, an example to inspire cities to explore waste treatment that is beautiful as well as economical.

Irrigation is used routinely to maintain lawns and trees in the city, but as water shortages increase, the city must explore a more waterconserving and drought-tolerant landscape. The landscape tradition that arose in the urban civilizations of the arid and semiarid regions surrounding the Mediterranean offers many models for the modern city, for example, the protected courtyard garden or patio. The courtyards nurture lush vegetation with minimal irrigation by protecting plants from dehydrating winds and radiated heat; the barren streets of the city heighten the aesthetic relief of the courtyards. The garden art of the Mediterranean and the Middle East also exploits the many physical properties and aesthetic qualities of water with great economy. A Persian garden accomplishes a great emotional and aesthetic effect with only a trickle of water.

The subtle, refined, and profound treatment of water in the Hispano-Islamic garden makes a 100 foot jet of water elsewhere seem a vulgar display of power. An art that developed over the course of thousands of years and spread with the Moslem religion west across North Africa to Spain and east to Pakistan and India, the Islamic garden takes many forms. Each form, however, reflects the inspired manipulation of water, employing the sight and sound of water to engender a cool atmosphere of serenity and retreat. Water cascades down sculpted channels or through plain runnels into brimming basins.

Slight variations in the shape of the channel produce wave patterns that catch the light in diverse ways. Water may appear precious, like a gem, as it flows over blue tiles. Water may bubble up from below the surface, or trace a graceful arc, or flow as a sheet over a molded edge. Water-poor cities should conserve their water by reserving irrigation for special or symbolic places or protected spaces where plants require minimal water. The importance of these places will thereby be heightened. Paley Park owes much of its success as an urban retreat to the contrast between its environment and the noisy, hot, dry city surrounding it.

The design for Foothill College, in the semiarid climate of Los Altos, California, as originally conceived, created an oasis garden to exploit the aesthetic impact of the contrast between irrigated and nonirrigated landscape. The architects designed the college as a compound of buildings, surrounding a central courtyard, on a hilltop, with parking below. The courtyard was designed as an oasis garden with lush vegetation sustained by irrigation; the hillside was seeded with drought-tolerant grasses. The contrast between the dry, brown hillside and the green, protected courtyard lent to the interior an atmosphere of comfort, retreat, and renewal. Since the college began to irrigate the hillside also, however, this atmosphere has been largely lost. It may be recaptured when water shortages in Northern California force the college to reduce irrigation.

In cities of a temperate, humid climate, enough rain falls to support a diverse community of plants without irrigation, so long as that water is permitted to infiltrate the soil and plants are protected from winds and radiant heat. Chestnut Park in downtown Philadelphia is paved and landscaped with plants native to that region. Rain falling within the park seeps between cracks in the pavement to the soil below. A deep layer of gravel beneath the topsoil serves both as a drainage device and as a reservoir, storing the water until plant roots can absorb it and preventing roots from becoming waterlogged. The plants have flourished and require no irrigation. Meanwhile, the park contributes no stormwater runoff to the city's sewers.

Designing to Conserve and Restore Water and to Prevent Floods

The prevention of floods and the conservation and restoration of water will only be accomplished by the cumulative effect of many individual actions throughout the city. But the impact of each will be insignificant, and might even be counterproductive, if not part of a comprehensive plan that takes into account the hydrologic system of the entire city and its region. Water pollution or flooding problems at one place may be generated somewhere else, and a solution to the water supply problem may, in the end, aggravate water pollution. The most effective, efficient, and economical solutions to urban water problems are frequently found upstream of where the problem is felt most forcefully.

Figure : *A weir was built on the Humber River (Ontario) to prevent a recurrence of a catastrophic flood.*

The Charles River watershed is the most densely populated river basin in New England. Its headwaters are sparsely developed, but the cities of Boston and Cambridge crowd the banks of its lower basin. The U.S. Army Corps of Engineers, in a 1965 flood control study of the Charles River watershed, concluded that a new dam must be built across the mouth of the Charles River to control flooding from urban runoff in the lower basin and that over the next thirty to forty years flood-control measures upstream must be taken to prevent flooding in the lower basin. They estimated that upstream flood-control structures would cost $100 million and, instead, recommended an action requiring one-tenth the cost:

The flood control management plan recommended by this Corps' study calls for federal acquisition and perpetual protection of seventeen crucial natural valley storage areas totalling some 8,500 acres. The logic of the scheme is compelling. Nature has already provided the least cost solution to future flooding in the form of extensive wetlands which moderate extreme highs and lows in stream flow. Rather than attempt to improve on this natural protection mechanism, it is both prudent and economical to leave the hydrologic regime established over the millenia undisturbed. In the opinion of the study team, construction of any of the most likely alternatives, a 55,000 acre/foot reservoir, or extensive walls or dikes, can add nothing.

The effective role of the wetlands in flood prevention was demonstrated while the Corps of Engineers was engaged in its study. In 1968 a large storm hit Boston, and urban runoff in the lower basin crested at the old Charles River Dam within hours. The upstream peak took four days to reach the dam. The wetlands in the headwaters filled with water, gradually releasing it over the course of a month. One stretch of the river widened from fifty feet to nearly a mile. Boston's second circumferential interstate highway was under construction at the time, and because rapid urbanisation threatened the wetlands, the Corps decided that acquisition of the wetlands was the most effective method of preserving their flood storage capacity. They selected seventeen natural storage areas, ranging in size from 118 to 2,340 acres, from among 20,000 acres of wetlands in the middle and upper reaches of the Charles River. In 1974, Congress approved and appropriated $10 million to buy the wetlands for nonstructural flood control. The Corps of Engineers made the first purchase in 1977. It will retain ownership of the land, and the Massachusetts Fisheries and Wildlife Division will manage the areas as wildlife refuges.

Denver, Colorado, is an outstanding example of a city that has implemented a comprehensive, coordinated set of strategies for managing its water. The devastating property losses caused by the 1965 flood provided the incentive for the formation of the Urban Drainage and Flood Control District in 1969. Earlier, each of the region's thirty-four independent local governments had employed different methods for calculating flood risks and for designing the capacity of their storm drainage systems. Some had designed storm drainage systems to accommodate a fifty year storm; others had provided for floods from a two year storm. The Urban Drainage and Flood Control District now works with local governments to insure the adoption and implementation of adequate and consistent floodplain regulations and to undertake master plans for individual watersheds. The *Urban Storm Drainage Criteria Manual*, published in 1969, guides the work in the district and insures consistent, state-ofthe-art drainage and flood control across the entire metropolitan region. The manual covers issues of policy, law, and planning related to flood control and storm drainage, the calculation of stormwater runoff, the design of the storm drainage system, and the mitigation of flood damage.

Each year Denver's Urban Drainage and Flood Control District compiles a list for master planning of between five and ten projects for which district aid has been requested by local governments. The project must be multijurisdictional, and local governments must agree to pay half of the cost of the study and half the cost of construction, and to assume ownership after completion. The district maps the one hundred year floodplain, prepares an outline of the work to be done, and coordinates consulting engineers on behalf of the local governments. The studies cover an entire drainage basin, rather than piecemeal projects. The master plan spells out where flood problems exist and recommends remedial measures. Its recommendations will include the adoption of floodplain regulations and the implementation of such projects as stormwater detention, channel improvements, and check dams along streams to create ponds and slow stream flow. The city and county of Denver now require property owners to pay a storm drainage service charge to help finance the construction and maintenance of the stormwater system. The amount of building and paved surface on the property determines the rate billed. In 1981, when the service charge was enacted, the city estimated that annual revenues would amount to $4.7 million.

Citizens of Denver have transformed a ten mile stretch of the South Platte River, which flows through downtown Denver, from a rubble-strewn, filthy, open sewer, lined by garbage and derelict land, into a landscaped park for water sports, public gatherings, bicycling and hiking, and nature study. Like the Urban Drainage and Flood Control District, the development of Denver's South Platte "Greenway" has its roots in the disastrous flood of 1965. A flurry of investigations and reports followed the 1965 flood, but little was done about the South Platte itself until a flood in 1973, an election year, brought the issue of the river and flood hazard under the public eye again. A nine-member task force, the Platte River Development Committee, appointed by Denver's mayor and backed by over $2 million in seed money from the city, proceeded to lay plans for the river, raise additional money from public and private sources, and implement park projects.

The Platte River "Greenway" now links eighteen parks with fifteen miles of interconnected trails; with 450 acres, it is Denver's largest single park. When complete, the "Greenway" will extend twenty-five miles upstream to the foot of the Rocky Mountains and twenty miles downstream to a state recreation area on the Colorado plains. Proponents hope that suburban communities will develop trails along the Platte's tributaries, so that eventually 120 miles of continuous river trails would lace the metropolitan region. The entire ten mile Platte River "Greenway" is now a regional centre for boating, lined by bicycle and hiking trails, and punctuated by parks. Check dams in the South Platte were designed to create white water "staircases" for canoes, kayaks, and rafts. Competitions are now held along the man-made "Challenge Run" and slalom kayak course. At one spot, where an eight-foot dam needed to retain water for a power plant made the river impassable by boat, a boat chute was created to permit boats and rafts to negotiate the dam without portage and to serve simultaneously as a flood control device. Carefully arranged weirs and rocks were placed to create a series of pools, riffles, and eddies ideal for recreational boating. The central channel of the Platte was excavated and large boulders and rocks placed to create a deeper stream during periods of low river flow. Water is now released from the upstream Chatfield Dam, a major flood control facility, in "recreation slugs" timed to enhance river flow for water sports during peak weekend recreation periods.

The many new parks along the Platte provide places to launch boats and to watch their progress through the chutes and slalom run. The Platte River Development Committee built the first park along the "Greenway" at the confluence of Cherry Creek and the South Platte River, where the city of Denver was originally founded. The large, terraced plaza of Confluence Park steps down to the river and provides an overview of part of the slalom run. Engineers designed the shape of the plaza and the opposite bank with a smooth profile to present minimal resistance to floodwaters, and designed the foundation to resist the river's hydrodynamic forces, by laying it directly on the riverbed and securing it with piles to the underlying shale bedrock. Years of accumulated rubble, which had blocked floodwaters and increased flood depth, were used in the construction of paths, boat chutes, and bank improvements.

An amphitheater across the river from Confluence Park was created with fill from river debris and the ruins of a bridge demolished by the 1973 flood. Pedestrian bridges, which link Confluence Park with the amphitheater and opposite banks in other parts of the "Greenway," are designed to pose no obstruction to floodwaters, since a major cause of past flood damage was the piling-up of debris at bridges in dams which diverted flóodwaters into adjacent parts of the city. The wooden pedestrian bridges are designed to come loose from their concrete piers when floodwaters reach the bridge deck. Cables attached to the bridge will hold it against the downstream bank until flooding subsides. All of the parks along the floodway are designed not only to resist flood damage, but also to provide flood storage. The grading for a new bicycle path at Centennial Park, for example, was based on flood hydraulics.

With increased use of the river for walking, bicycling, and boating has come a heightened awareness of the river's water quality and a strong constituency for improving and maintaining that quality. Many sources of water pollution have been removed from the river banks as a consequence: a dump has been converted to a nature preserve; a highway maintenance yard piled with salt and sand has become Frog Hollow Park. Pressure has been brought upon the city to cease dumping street sweepings and salt-laden snow in the river. The residential neighbourhoods bordering the South Platte, several of them Denver's poorest, have gained new parks and a river environment free from former nuisances and hazards.

The Platte River "Greenway" was accomplished through the coordinated efforts of public and private organisations and individual citizens. The Platte River Greenway Foundation, established as a nonprofit, tax-exempt institution, ultimately collected over $6 million from federal, state, and local governments, from private foundations, and from individuals. The foundation, though private, cooperated closely with the city from its inception; funded and coordinated the implementation of projects on behalf of the city; and then turned over the responsibility of maintenance to the city's parks department.

Rooftops, plazas, and parking lots often provide the only space to detain stormwater in densely built downtown areas, and Denver is no exception. The city of Denver requires new and renovated buildings in the Skyline Urban Renewal District to detain stormwater on site. The alternative, upgrading the existing storm sewer system to accommodate the increased runoff, would have been prohibitively expensive and would have increased flooding in the nearby South Platte River. Developers have used a combination of rooftops, plazas, and parking lots to detain stormwater. Roofs in the Denver area are designed to support a snow load equivalent to approximately six inches of water. Engineers designed a "detention ring" to fit around the drain of a flat roof, which ponds up to three inches of water, then releases it at the rate of one-half inch per hour. A safety feature permits a severe storm to overflow the ring. Denver-area plazas and parking lots have been designed to store stormwater runoff with minimal inconvenience to pedestrians. One depressed, downtown Denver plaza, constructed above three floors of underground parking, accommodates runoff from the ten year storm; stormwater drains directly to the sewer at the rate of one inch per hour. Ponding does not disrupt use of the plaza, since elevated portions of the plaza permit pedestrians to walk across it when lower portions are flooded.

Existing building codes in most American cities require that roofs be designed to withstand the equivalent of six inches of water over a short period (usually twenty-four hours), and a few cities have incorporated rooftop detention of stormwater into building codes. European cities like Stuttgart have applied the use of "wet roofs" to reduce the building's heat load as well, and thus decrease energy consumption for air conditioning. If incorporated into roof garden design, stormwater detention can also become an aesthetic amenity.

In little more than one decade, Denver has achieved considerable success in reclaiming its waters. Consider how much might be accomplished in the construction of a new city unhampered by existing buildings, streets, and drainage systems. Such a case is the new town of Woodlands, Texas, with a projected ultimate population of 150,000. When developer George Mitchell first decided to build a new town on 20,000 acres of pine-oak woodland north of Houston, he envisioned a city that would spring up in the midst of the woods, in harmony with the forces of nature. He formed the Mitchell Energy and Development Corporation and hired an interdisciplinary team of planners, engineers, scientists, and market specialists. Initially this team consisted of four firms. Over the following decade the team was expanded to include a well-staffed corporation with dozens of consultants.

By 1971, when the preliminary ecological planning study and parallel market research were complete and a general plan for the new town was underway, water had emerged as the critical factor. The Woodlands' "natural drainage system" exploits the capacity of natural, wooded floodplains to accommodate stormwater runoff and of well-drained soils to soak up and store rainfall. It reduces the combination of increased flooding and lower stream flows normally associated with urbanisation, it maintains water quality, and recharges the aquifer that underlies neighbouring Houston. The wooded floodplain, drainage channels, and recharge soils form a townwide open-space system, a natural drainage system that represents a substantial savings over the cost of constructing a conventional storm sewer system. When it was originally proposed, engineers compared the cost of the natural drainage system to that for a conventional storm system and estimated that the natural drainage system would save the developer over $14 million.

Much of the Woodlands site is very flat, with extensive areas of poorly drained soils. The construction of a traditional storm drainage system would have entailed clearance of extensive woodlands, and lowered water tables with loss of trees. It would also have increased flooding and degraded water quality downstream and, combined with an estimated 15 million gallons per day withdrawal from underlying aquifers, might have contributed to further ground subsidence under the city of Houston. The firm of Wallace McHarg Roberts and Todd, landscape architects and ecological planners, conceived a natural drainage system to resolve these problems and enable the developer to retain his vision of the future city.

The natural drainage system is composed of two subsystems: one stores and absorbs rainfall from frequent storms; the other drains floodwater from major storms. The general plan responded to the major drainage system by locating large roads and dense development on ridge lines and higher elevations, while preserving the floodplains in parks and open land, and allocating lowdensity housing to the intermediate area. Use of the floodplains and drainage channels as open space works well from both ecological and social standpoints. Most of the spectacular trees on the site occur within the floodplains of two major creeks large, evergreen magnolias, water and willow oaks, and towering pines. These same floodplains also harbour a diverse, abundant native wildlife, including white-tailed deer, opossum, armadillos, bobcats, and many birds, and provide the corridors along which they move. The wooded easements required for drainage and flood control purposes are in most cases sufficient to insure that all but the most sensitive wildlife species may remain.

A continuous system of hiking, equestrian, and bicycle trails runs along the drainage network, linking all parts of the new town. Although this larger floodplain network drains runoff from major storms, well-drained soils and ponds absorb or store rain close to where it falls, either in private yards or in nearby parks. This local drainage system responds to subtle changes in topography and soils. Roads, golf courses, and parks are designed to impound stormwater and enhance its absorption by well-drained soils. Maintaining the structure of these soils, so essential to their ability to absorb water, required strict regulation of construction activities. Areas designated as "recharge soils" were left wooded. In some cases, building construction proceeded within a fenced-off zone that extended only a few feet on all sides from the building foundation. This practice has produced a new town with the appearance of having literally sprung up within the woods.

Models estimating the increase in peak flows generated by development at the Woodlands revealed that they will increase by only 55 percent, as compared to the 180 percent increase resulting from current "normal" development in Houston. Studies indicate that the water quality of urban runoff in the phase one portion of the new town is much better than that of other Houston residential areas. The final test of the natural drainage system occurred when a record storm hit the area in April 1979. Nine inches of rain fell within five hours, and no house within the Woodlands flooded although adjacent subdivisions were awash.

The economic benefits of a natural drainage system may not elsewhere be as dramatic as in Woodlands with its extensive flat areas and poorly drained soils, but they will accrue nonetheless. The Woodlands is and will continue to be a showpiece of drainage design, from the most mundane details of pavement and channel design to the coordination of soils, ponds, swales, and floodplains into a comprehensive drainage system.

A Plan for Every City

The successful management of water in the city will require comprehensive efforts, many individual actions, and the perception that storm drainage, flood control, water supply, water conservation, waste disposal, and sewage treatment are all facets of a much broader system. Every city should construct a framework within which both the consequences of major metropolitan efforts and the cumulative effect of individual actions can be appreciated.

The flow of water into and through the city including where it comes from, how and where it is used, treated, and released, and the seasonal variation of this pattern varies from city to city, depending on regional climate, topographic setting, pollution sources, and urban form. Do floods threaten a major portion of the city; and is development upstream the greater problem or constriction of the floodplain within the city? Is the city's water supply threatened by pollution of ground water or surface water, or by competing demands with other towns and cities in the region? Are large, industrial pollutors the problem, or combined sanitary and storm sewage overflows? Are extremes of high flows and low flows a problem, or limited water circulation? Identifying the areas at most risk to flood hazard and those that currently provide flood storage will help in devising a comprehensive flood control strategy.

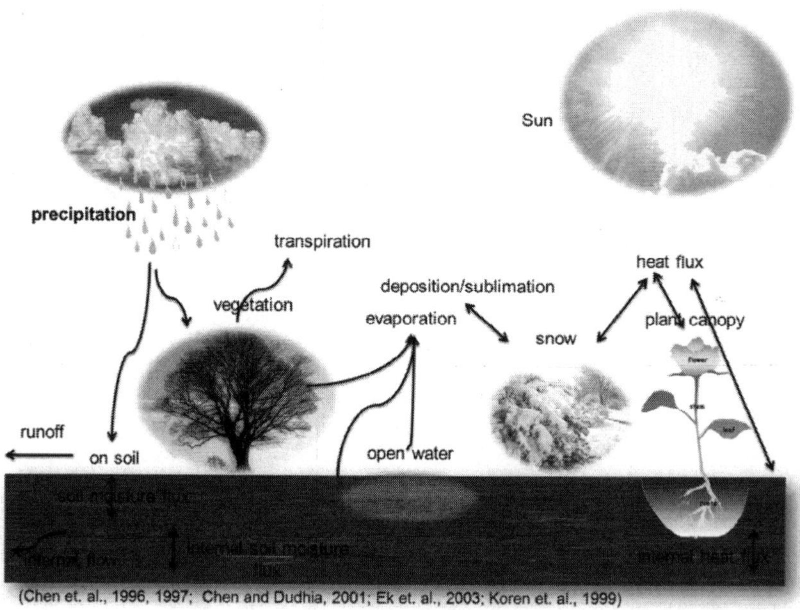

Figure : *Many different process lead to movements and phase changes in water.*

Identifying the major sources of water pollution within the city, the dispersion patterns of pollutants within surface and ground water, and water bodies with poor circulation will aid in singling out the most severely contaminated places. Knowing the most significant water resources, those that currently provide the city with water or have the potential to do so in the future, and the areas that are most sensitive to water pollution, like aquifer recharge areas, headwater streams, lakes, and ponds, will help to preserve those resources. A comprehensive plan to prevent floods and conserve and restore the city's water should:

- address the city's most critical problems of flooding, water pollution, and water supply, with particular attention to reducing risk in the most floodprone or contaminated areas
- protect the city's most important water resources, both those

currently used for water supply and those with potential to satisfy increasing demand

- locate new parks and other landscaped open space to preserve flood storage in the headwaters and the floodplain downstream, and to enhance recharge of ground water
- encourage new industry, waste disposal sites, and other polluting land uses to locate outside floodplains and ground water recharge areas that are highly vulnerable to water pollution
- locate new public buildings outside flood-prone areas and encourage new residential and commercial de velopment to do the same
- provide a plan for relocation and reconstruction after a major flood
- explore settlement patterns that would facilitate the reuse of wastewater after treatment
- exploit the flood protection and water-restoring abilities of existing wetlands
- increase the visibility of water in the city as well as public access to it.

Every new building, street, parking lot, and park within the city should be designed to prevent or mitigate flooding and to conserve and restore water resources. Every project should:

- address the relationship between the project's site and the city's critical flooding, water pollution, and water supply problems, as well as specific hazards and resources that exist on the site and in its immediate neighbourhood
- site and design buildings and landscaping to avoid flood damage
- exploit the ability of rooftops, plazas, parking lots, and the earth to detain or retain stormwater runoff design parks in floodplains to store floodwater and withstand flood damage
- design the size, depth, shape, and shoreline of urban water bodies to enhance water circulation and store stormwater
- select hardy plants that require little, if any, irrigation, fertilizers, or pesticides, and protect plants from desiccating winds
- utilise stormwater, if sufficiently uncontaminated by salts and pollutants, or treated wastewater to meet plant water needs
- exploit the aesthetic properties of water without wasting it.

Water problems and their severity vary from city to city, but every city must manage its own water resources. Cities of the past and present plagued by water problems have pioneered solutions to flood control, water conservation, and water restoration. Many of these models are applicable to every city, not just those with semiarid climates or with intensely developed floodplains. Opportunities for preventing floods, for preserving water quality, and for conserving water exist in the design of every new building and park as well as every metropolitan plan, at the centre of downtown and at the urbanising metropolitan fringe. When the urban water crisis comes, it will probably hit fast-growing cities in and regions first, but it will extend inevitably to cities in humid regions as well. Eventually, every city will have to design a comprehensive plan for water management, including the regulation of urban form and density in headwaters and floodplains, the regulation of water use, with attendant implications for landscape design, and the careful siting of waste disposal sites and industrial and sewage outfalls.

The knowledge for such a plan exists today; the underevaluation of water is the major obstacle to its implementation. Once the water crisis forces cities to charge full value for water, the support for water conservation will follow. When cheap water is a thing of the past, rainfall will be cherished, runoff utilised, and flooding reduced. Cities will protect their water from contamination and reuse it after treatment. City parks and private grounds will acquire a drought-tolerant landscape. The use of water in public spaces will be restrained, but the impact will be powerful.

In the next decade, the dilapidated, outmoded water supply, wastewater treatment, and storm drainage systems in many older American cities will have to be overhauled. This will entail the expenditure of billions of dollars and considerable upheaval in dense urban centres. Short-term expediency must not prevail; the opportunity for redesign must be seised.

2

The Groundwater Resource

Groundwater may be defined as subsurface water that occurs beneath the water table in soils and geologic forms that are fully saturated. It is an integral part of the hydrologic cycle, and any approach to groundwater problems should recognise this. Groundwater is not only an important natural resource but also an essential part of the natural environment. Much of the folklore and many of the widely held misconceptions about the nature of groundwater. For example, that it occurs in underground lakes, rivers, and veins and can be detected by listening for the noise are dispelled by Lehr and Pettyjohn (1975) in their description of groundwater and its flow patterns.

Figure: *Shipot, a common source of drinking water in a Ukrainian village.*

The global hydrologic cycle consists of the movement of water between the oceans and other surface water, the atmosphere, and the land. A part of the precipitation, either as rainfall or melting snow, infiltrates the ground and percolates down through the unsaturated soil, known 18 as the aerated zone, to the zone of saturation or the water-table level. An estimate of the water balance of the world, shows the relative volumes of water contained in each part of the hydrologic cycle. Only 2.7% of all the water on the planet is fresh, and of that only 0.36% is easily available to users.

It has been estimated that about 40,000 billion gallons per day (bgd) pass over the conterminous United States as water vapour. Approximately 10% of this, 4,200 bgd, is precipitated as rain, snow, sleet, or hail, equivalent to an average uniform annual rainfall of 30 inches nationwide. About two-thirds of the precipitation evapourates or is transpired by vegetation. The remaining 1,450 bgd accumulates in ground and surface waters, flows to the sea or across national boundaries, is consumed, or evaporates from reservoirs.

Table: *Estimate of the Water Balance of the World*

Parameter	*Surface Area (km)x10*	*Volume (km)x10*	*Volume (%)*	*Equivalent Depth(m)**	*Residence Time*
Oceans and Seas	361	1370	94	2500	~4000 yrs
Lakes and Reservoirs	1.55	0.13	<0.01	0.25	~10 yrs
Swamps	<0.1	<0.01	<0.01	0.007	1-10 yrs
River Channels	<0.1	<0.01	<0.01	0.003	~2 wks
Soil Moisture	130	0.07	<0.01	0.13	2 weeks-1 yr
Ground water	130	60	4	120	2 wks-10,000 yrs
Icecaps and Glaciers	17.8	30	2	60	10-1000 yrs
Atmospheric Water	504	0.01	<0.01	0.025	~10 days
Biospheric Water	<0.1	<0.01	<0.01	0.001	~1 wk

* Computed as though storage were uniformly distributed over the entire surface of the earth

Source: R. Allan Freeze, John A. Cherry. *Groundwater* (Engelwood Cliffs, N.J.: Prentice Hall, Inc., 1979), p. 84. Reprinted by permission of the publisher.

Of the 1,450 bgd only 675 bgd are usually available for intensive beneficial uses.

The ability of an aquifer to store and transmit water is a function of its permeability and porosity. When the saturated substratum is sufficiently permeable to store and transmit significant quantities of water, the geological formation is called an aquifer. There are two main types of aquifers, confined and unconfined. An unconfined or water-table aquifer contains water under atmospheric pressure; the upper 19 surface of the water is called the water table and may rise and fall according to the volume of water stored, which is dependent upon seasonal cycles of natural recharge. A perched aquifer is one in which a limited layer, or lens, of impermeable material occurs above the water table, forming a thin zone of saturation above it. It is a type of water-table aquifer. The second major category of aquifer is the confined, or artesian, aquifer.

These are bounded top and bottom by layers of relatively impermeable geologic formations termed aquitards, or confining layers. The aquifer is completely saturated with water that is under greater than atmospheric pressure. An artesian aquifer is not recharged everywhere uniformly, but in one or more general recharge areas. Water levels in nonpumping wells of unconfined aquifers correspond to the level of the water table and therefore vary according to the volume of water in storage. Water levels in nonpumping wells tapping confined aquifers are dependent upon the artesian pressure of the water in that aquifer, and in some cases the water may exceed the top of the well casing, thus causing a flowing well. The hypothetical projection of such water levels is known as the potentiometric surface. The aquitards, or confining layers, are not totally impermeable and permit some recharge or discharge to lower confined aquifers. Aquifers can occur in unconsolidated materials such as sand and gravel or in consolidated material or bedrock. The latter may consist of carbonate rock, volcanic rock, or fractured igneous, metamorphic, and sedimentary rocks. Sand and gravel aquifers usually contain the most groundwater, but high-yield wells can also occur in carbonate and volcanic rocks.

Groundwater Movement

Groundwater moves in response to gravity, pressure, and friction, the first two driving the water, the latter resisting motion. Due to the complexity of the channels through which groundwater flows, it is difficult to construct a model of the movement of groundwater at a microscopic level. The French engineer Darcy, however, formulated an empirical law in the mid-nineteenth century that effectively averages the microscopic complexities, providing a macroscopic model for groundwater movement. Darcy's Law relates the rate at which groundwater flows across a surface to the rate of change of energy of the groundwater along the flow path. Under ideal 20 homogeneous isotropic geologic groundwater conditions, the average linear groundwater velocity ((v)) can be expressed by the Darcy relation:

$$v = -[K\ dh/dl]/n$$

where:

v = the average linear groundwater velocity

K = hydraulic conductivity

dh/dl = hydraulic gradient

n = transport porosity

For both granular and fractured media, the same equation is used to compute values for average linear groundwater velocity. The great difference is in the value of *n*. For granular media n nearly represents the total porosity; in fractured rock or clay n represents the total void space in connected fractures within a unit volume of media. The value of n in granular material is usually between 0.3 and 0.45; in fractured material n is usually very small, of the order of 10-*f*"2 or 10-*f*"4. When the values of *K* and *dh/dl* are similar in granular and fractured media, the average linear groundwater velocity may be orders of magnitude larger in fractured media than in granular media. The velocity of groundwater in a horizontal sand and gravel aquifer is usually in the range of 0.05 to 1 meter per day.

Darcy's Law is central to the derivation of equations used to model the flow of groundwater. Mathematical models, based on the physics of groundwater flow and boundary conditions imposed by the groundwater basin in question, usually take the form of a boundary-value problem. Bear (1972) discusses the application of the three methods used for solving such boundary-value problems, namely analytical, analogue, and numerical. Analytical methods result in explicit mathematical expressions of the solution and are useful for problems where the governing equations may be simplified and the boundaries of the basins idealised. Analogue models are scaled physical models usually involving the construction of electronic equipment to simulate groundwater flow. Numerical mathematical techniques are the basis for digital computer simulation of transient groundwater flow in aquifers. Digital simulation of groundwater flow requires an expertise in computer programming and its use is bound to digital computers. However, it is more flexible in its ability to handle irregular groundwater boundaries and variations of groundwater movement through time and space than are the other two modelling techniques. The two most widely used numerical methods for solving groundwater equations are the finite-difference and finite-element techniques.

The main difference between groundwater and surface water is that the movement of groundwater takes place very slowly. Mixing of groundwater is slow, in contrast to the good mixing potential in the turbulent flow of most surface water. This factor is extremely important when considering the fate of contaminants.

Occurrence and Natural Quality of Groundwater

Most aquifers occur within 2,500 feet of the surface of the land. They may be thick or thin, extensive or local, very near the land surface or at considerable depths. It has been estimated that 30% of the stream flow of the United States is supplied by groundwater that emerges as natural springs or other seepage. In certain areas in times of drought, the stream flow in lowflow months may be largely provided by groundwater. The interrelatedness of surface water and groundwater is further underlined by the fact that under certain conditions seepage from lakes, rivers, streams, reservoirs, and canals may recharge aquifers.

The quantity of groundwater in storage is much greater than the volume of surface water available in streams and lakes. It comprises more than 96% of all the fresh water in the United States, the remaining 4% occurring in lakes, rivers, and streams. Estimates of groundwater within 2,500 feet of the land surface in the conterminous United States range from 33 quadrillion to 59 quadrillion gallons to as much as 100 quadrillion gallons. As a point of comparison, Lake Michigan contains 1.3 quadrillion gallons of water. Not all of the groundwater in storage is available for use, however, because some is bound to soil particles. The cost of extraction may also restrict its use in certain areas. It would be possible to drill a well with a yield sufficient for domestic use in almost every region of the country. The country has been divided up into 10 groundwater regions. This is considered the best broad classification of the groundwater situation in the conterminous United States. The division is based on the types of aquifers.

1. The western-mountain ranges consist mainly of igneous, metamorphic, and consolidated sedimentary rocks. Most of the groundwater occurs in rock fractures. The large amount of precipitation that falls in this area recharges aquifers in adjacent regions.
2. Alluvial basins consist of valleys surrounded by mountains in the arid southwest. Water levels in many aquifers have declined

over recent years due to the use of large volumes of groundwater for agricultural irrigation.

3. The Columbia Lava Plateau consists of lava flows and unconsolidated sediments. In some areas there are large supplies of groundwater.
4. The Colorado Plateau has high, dry plateaus of sedimentary shale and sandstone and has a scarcity of productive aquifers.
5. The High Plains are extensive and semi-arid to sub-humid. The unconsolidated sedimentary rocks form the Ogallala Aquifer. Due to extensive agricultural irrigation, withdrawals have greatly exceeded groundwater recharge in some areas.
6. The glaciated central lowlands consist of glacial drift, sand, and gravel forming major groundwater reservoirs. In some heavily populated or highly industrialised areas the groundwater has been overdeveloped or polluted.
7. The unglaciated Central Region is composed of horizontal consolidated sedimentary rocks, with limestone and sandstone formations providing the major aquifers. The yields are low to moderate and adequate for domestic supply but would not be sufficient for irrigated agriculture.
8. The unglaciated Appalachians have high-yielding limestone and sandstone aquifers, and lower yielding shale aquifers. The water table may occur at a considerable depth below the surface. This region has an abundance of surface water year round.
9. The glaciated Appalachians region is similar to the unglaciated Appalachians.
10. The Atlantic and Gulf Coast Plain has an abundance of groundwater and surface water. The geology consists of unconsolidated gravel, sand, silt and clay, and limestone. Supplies of groundwater may not be abundant in some areas of Texas, Louisiana, Mississippi, and Alabama. Florida has a prolific aquifer.

Unfortunately there is more than one classification system for dividing the country up into water regions. For the purpose of compiling and analysing water-resources data for both surface water and groundwater, the U.S. Water Resources Council divided the nation into 21 major water-resources regions, 18 within the conterminous United States, and the other 3 being Alaska, Hawaii, and the Caribbean area. These 21 regions are further subdivided into 106 subregions.

The regional divisions are areas that contain either the drainage area of a major river or the combined drainage of a series of rivers.

The EPA used another classification system in compiling data on groundwater pollution for various regions of the United States. The EPA system most closely resembles that described by Lehr et al. (1976), which is summarised above. Along with the geological characteristics of the groundwater regions, climatic conditions in particular, precipitation and temperature have effects upon the quantity and quality of the groundwater available. Variations from average precipitation may cause droughts or floods. Water is lost from the land to the atmosphere by evaporation and transpiration, or evapotranspiration. When evapotranspiration exceeds precipitation, groundwater recharge by percolation does not occur. In the arid areas of the Southwest, for example, annual precipitation is less than 10 inches, and potential evaporation could exceed this 4 to 20 times. In such areas groundwater recharge occurs mainly in wet, or multiyear, cycles. In the eastern states, the annual precipitation exceeds the evaporation, thus providing surpluses that contribute to stream flow. Regional variations in temperature may also affect groundwater in ways unrelated to surface evaporation. For example, frozen ground does not permit infiltration of rain, thus causing floods and preventing groundwater recharge.

The United States Geological Survey (USGS) published summary appraisals of the nation's groundwater resources between 1974 and 1982 for each of the regions that were briefly outlined above in the U.S. Water Resources Council scheme. These geological survey professional papers list groundwater usage for each region by categories of use (e.g., municipal water supply and agricultural irrigation). They list the natural groundwater quality and any problems that may be encountered due to the presence of natural iron in the water or to its salt content and hardness. In addition, they list present and potential anthropogenic threats to the quality of groundwater.

The National Water Summary 1984, also put together by the USGS, contains descriptions of the occurrence, use, and general quality of groundwater resources of each state. Each summary contains: (1) the physiographic, hydrologic, and geologic framework of the state's groundwater system; (2) a description of the principal aquifers; (3) groundwater withdrawals and water-level trends; (4) a description of the state's groundwater management program(s); (5) maps showing geographic distribution of the principal aquifers; figures illustrating the areal distribution of major groundwater withdrawals and trends in water levels; and selected references.

Table: *Summary Appraisals of the Nation's Groundwater Resources, by Region*

(A)	Ohio — R. M. Boyd, Jr., 1974.
(B)	Upper Mississippi — R. M. Boyd, Jr., 1975.
(C)	Upper Colorado — Don Price and Ted Arnow, 1974.
(D)	Rio Grande — S. W. West and W. L. Broadhurst, 1975.
(E)	California — H. E. Thomas and D. A. Phoenix, 1976.
(F)	Texas-Gulf — E. T. Baker, Jr. and J. R. Wall, 1976.
(G)	Great Basin — Thomas E. Eakin, Don Price, and J. R. Harrill, 1976.
(H)	Arkansas-White-Red — M. S. Bedinger and R. T. Sniegrocki, 1976.
(I)	Mid-Atlantic — Allen Sinnott and Elliott M. Cushing, 1978.
(J)	Great Lakes — William G. Weist, Jr., 1978.
(K)	Souris-Red-Rainy — Harold O. Reeder, 1978.
(L)	Tennessee — Ann Zurawski, 1978.
(M)	Hawaii — K. J. Takasaki, 1978.
(N)	Lower Mississippi — J. E. Terry and C. T. Bryant, 1979.
(O)	South Atlantic-Gulf — D. J. Cederstrom, E. H. Boswell, and G. R. Tarver, 1979.
(P)	Alaska — Chester Zenone and Gary S. Anderson, 1978.
(Q)	Missouri Basin — O. James Taylor, 1978.
(R)	Lower Colorado — E. S. Davidson, 1979.
(S)	Pacific Northwest — Bruce L. Foxworthy, 1979.
(T)	New England — Allen Sinnott, 1982.
(U)	Caribbean — Fernando Gómez-Gómez and James E. Heisel, 1980.

Source: U.S. Geological Survey Professional Paper 813. U.S. Government Printing Office.

The quality of groundwater is often described in terms of hardness and salinity. Hardness reflects its calcium and magnesium content and is usually expressed as the equivalent amount of calcium carbonate. It can be viewed as a measure of usefulness for domestic and industrial purposes.

Water quality, however, is usually defined in terms of the concentration of its chemical constituents. As water moves through the hydrologic cycle, it interacts with the atmosphere, soils, and subsurface geologic formations, all of which affect its chemical composition. Thus, there is a natural background level for the chemical content of the water, and this level varies regionally and may be subsequently augmented from industrial and domestic sources. Freeze and Cherry (1979) discuss the natural chemical content of groundwater in detail. Rainwater is saturated with oxygen, nitrogen, and carbon dioxide gases and is usually slightly acidic, having a pH of about 5.6. The acidity may be increased by industrial pollutants, namely, the oxides of sulfur and nitrogen. The more acid the rainwater, the more likely it is to react with the geologic materials with which it comes into contact.

Rainwater percolating through the soil may increase in acidity due to the biological processes that occur in that zone. Plant and microbial respiration produce carbon dioxide, which increases acidity. It would be possible for the percolating water to become supersaturated with carbon dioxide. Acidity may also be increased by products of decomposition such as humic and fulvic acids, nutrient uptake by roots, and nitrifying bacteria, but these are minor factors compared with the amount of carbon dioxide produced by respiration. Thus, the water undergoes a chemical change during its passage through the soil to the underlying water-bearing formations.

Table: *Simple Groundwater Classification Based on Total Dissolved Solids*

Category	***Total Dissolved Solids(mg/l or g/m)***
Fresh water	0-1000
Brackish water	100-10,000
Saline water	10,000-100,000
Brine water	More than 100,000

Source: Freeze and Cherry, 1979.

Table: *Classification Based on Dissolved Inorganic Constituents in Groundwater*

Major Constituents (greater than 5 mg/l)	
Bicarbonate	Silicon
Calcium	Sodium
Chloride	Sulfate
Magnesium	Carbonic acid
Minor Constituents (0.01-10.0 mg/l)	

Boron	Nitrate
Carbonate	Potassium
Fluoride	Strontium
Iron	
Trace Constituents (less than 0.1 mg/l)	
Aluminum	Molybdenum
Antimony	Nickel
Arsenic	Niobium
Barium	Phosphate
Beryllium	Platinum
Bismuth	Radium
Bromide	Rubidium
Cadmium	Ruthenium
Cerium	Scandium
Cesium	Selenium
Chromium	Silver
Cobalt	Thallium
Copper	Thorium
Gallium	Tin
Germanium	Titanium
Gold	Tungsten
Indium	Uranium
Iodide	Vanadium
Lanthanum	Ytterbium
Lead	Yttrium
Lithium	Zinc
Manganese	Zirconium

Source: S. N. Davis and R. J. M. DeWiest, Hydrogeology (New York: John Wiley and Sons, Inc., 1966). Reprinted by permission of the publisher.

In the saturated zone, the water reacts with the geologic formations, increasing the content of total dissolved solids. Thus, the chemical quality of groundwater depends both upon its age and the geological formations encountered in its flow history. Carbonate formations would increase the magnesium and calcium content and also the concentration of bicarbonate due to the dissolution of calcite ($CaCO_3$) and dolomite ($MgòCaCO_3$). Aluminosilicate minerals would increase the concentrations of sodium, potassium, magnesium, calcium, and silicon hydroxide ($Si(OH)_4$) in the water.

In consolidated deposits consisting of minerals from a range of sedimentary, igneous, and metamorphic sources, the order of encounter of the groundwater with the different assemblages determines the chemical constituency at a particular point in time. Sulfate-bearing minerals such as gypsum and anhydrite, although they occur less frequently than the carbonate and crystalline formations, are characterised by high solubilities. Thus, in older groundwater that has encountered sulfur-bearing minerals, the sulfate anion may dominate the bicarbonate anion. In very deep, old groundwater, the chloride anion may dominate both the bicarbonate and sulfate ones due to the presence of readily soluble minerals such as halite (NaCl) and sylvite (KCl).

The U.S. Public Health Service (USPHS) standard for total dissolved solids (TDS) in drinking water is 500 ppm, although the 1974 Safe Drinking Water Act considers waters containing up to 10,000 ppm TDS as potential sources of drinking water. The four types of naturally occurring groundwater that often exceed 10,000 ppm TDS are connate water, intruded seawater, magmatic and geothermal water, and water affected by salt leaching and the products of evapotranspiration. In many areas of the United States, groundwater can be used for drinking with no pretreatment, in which case it is said to be a raw resource. In other areas pretreatment to correct hardness or colour may be necessary. Where contamination by pathogenic organisms is a potential threat, chlorination may be indicated. Water from some aquifers is unusable due to its salinity or to the presence of naturally occurring toxic substances such as arsenic or radionuclides. Most of the freshwater aquifers are underlain by brackish or saline aquifers, and in general, salinity may be said to increase with depth.

Use of Groundwater

Groundwater is a major natural resource in the United States and is often more easily available than surface water. It is estimated that more than 50% of the population uses groundwater as its primary source of drinking water. These estimates are from unpublished data for 1970 from the USGS. They indicate that groundwater delivered by community systems supplied 29% of the population, and an additional 19% had its own domestic wells. Approximately 36% of the municipal public drinking-water supplies came from groundwater. Of the rural population, 95% was dependent upon groundwater for drinking purposes in 1970. Of the total groundwater usage, 60% was used for public supplies and 40% for rural supplies. EPA's National

Statistical Assessment of Rural Water Conditions estimated that 72.3% of all major rural household supplies was extracted from groundwater. This dependence varied regionally, with 87.5% of north-central rural households and 56.7% of western households dependent upon groundwater for water supplies. This table indicates that 97% of water withdrawn for the nation's domestic rural use is supplied by groundwater. It is clear that the states west of the Mississippi River, in the area where irrigated agriculture is prevalent, depend heavily on groundwater. Arkansas, Nebraska, Colorado, and Kansas use more than 90% of their groundwater for agricultural activities. Irrigation was the principal use of all groundwater withdrawn across the country in 1980. The more humid eastern portion of the country is far less dependent on groundwater. The western states, again, are the main users of groundwater.

The most recent figures for groundwater use are from 1980 USGS data. The total use of fresh groundwater showed an increase of 22% from 1970 to 1975 and an increase of only 7% from 1975 to 1980. This change in the rate of increased use may be attributable to the fact that high demand for groundwater may decrease the ease and efficiency of its withdrawal, and in some cases its quality.

Table: *Groundwater Withdrawals as a Percentage of Total Freshwater Withdrawals for Rural Domestic Supply*

State	***Percentage***	***State***	***Percentage***
Alabama	100	Montana	94
Alaska	99	Nebraska	100
Arizona	100	Nevada	94
Arkansas	100	New Hampshire	98
California	93	New Jersey	100
Colorado	36	New Mexico	97
Connecticut	100	New York	89
Delaware	100	North Carolina	100
District of Columbia	0	North Dakota	100
Florida	100	Ohio	90
Georgia	100	Oklahoma	83
Hawaii	90	Oregon	87
Idaho	96	Pennsylvania	100
Illinois	97	Rhode Island	100
Indiana	90	South Carolina	100

Contd...

State	Percentage	State	Percentage
Iowa	100	South Dakota	94
Kansas	86	Tennessee	100
Kentucky	91	Texas	84
Louisiana	100	Utah	90
Maine	98	Vermont	85
Maryland	100	Virginia	100
Massachusetts	100	Washington	78
Michigan	100	West Virginia	95
Minnesota	100	Wisconsin	100
Mississippi	100	Wyoming	92
Missouri	74		
Total national percentage			*97*

Source: After Heath, 1985.

Escalated costs resulting from these factors and the cost of the water itself have perhaps influenced users, especially irrigators, to be more efficient with groundwater use. Actual water-use figures for 1975 and 1980 (including saline groundwater withdrawals and water used for the generation of thermoelectric power) are given.

Table: *Historical Trends in U.S. Groundwater Use as a Percentage of Withdrawals: 1950-1980*

	1950	1960	1970	1975	1980
Total Fresh Groundwater Withdrawals (bgd)	34	50	68	82	88
Public Supplies (%)	12	13	14	13	14
Rural Supplies (%)	8	6	5	5	5
Irrigation (%)	62	68	66	69	68
Industry (%)	18	13	15	14	13

Note: May not total 100% due to rounding.

Sources: Murray and Reeves, 1972, 1977; Makichan and Kammerer, 1961; Soiley et al.,1983.

In an average American community, the average per-capita water consumption is approximately 159 gallons per day (Last, 1980). In an American home, depending on the nature of the residence and the climate, exterior residential use may range from 5% to more than 150% of the interior use, averaging about 75%. This is primarily for watering lawns. Personal use of water, excluding laundry, toilet flushing, and so forth, accounts for 40% of the internal residential use and averages out to 6.3 gallons per capita per day.

The safe yield of a groundwater basin is defined as the amount of water that can be withdrawn annually without producing an undesired effect. Undesired results include depletion of the groundwater reserves, intrusion of water of an undesirable quality, contravention of water rights, the deterioration of the economic advantages of pumping, excessive depletion of stream flow, and land subsidence. Any withdrawal in excess of a safe yield is termed an overdraft. It has been suggested that the optimal yield would be a more useful concept and should be determined by an optimal groundwater-management scheme that includes and best meets economic and social objectives associated with the use of water. Under certain conditions, optimal yields may involve the mining of groundwater to depletion or complete conservation.

Increased demands on groundwater have strained the supply in certain regions, resulting in reduced artesian pressure, land subsidence, reduced spring and stream flow, and the intrusion of salt water.

Table: *U.S. Groundwater Use: 1975, 1980*

	1975	***1980***
Population served by groundwater for public supplies	64,700,000	73,700,000
Total groundwater withdrawn (mgd) for public supplies	11,000	12,000
Rural use of groundwater (mgd)		
Domestic	2,700	3,300
Livestock	1,200	1,200
Total withdrawn for irrigation (mgd) toirrigate a total of 54,000,000 acres (1975) and 63,125,000 acres(1980)	57,000	60,000
Groundwater withdrawals for industrial use (mgd)		
Fresh	11,000	10,000
Saline	980	930
Water used for generation of thermoelectric power	1,390	1,600
Total withdrawals of groundwater (mgd)	85,270	89,030

Sources: Murray and Reeves, 1977; Soiley et al., 1983.

This kind of groundwater overdraft is occurring in the High Plains from Texas to Nebraska, parts of California, Arizona, Louisiana, Florida, New Mexico, Arkansas, Wisconsin, Illinois, and North Carolina. It has been reported that groundwater levels of the Carrizo Aquifer in south-central Texas have declined as much as

***Table:** Allocation of Water Use In U.S. Communities*

Use	***Percentage***
Residential	40
Commercial	15
Industrial	25
Public	5
Unaccounted for	15
Total	100

Source: J. M. Last, in Maxcy-Rosenau, eds. *Public Health and Preventive Medicine*,11th ed. (Norwalk, Conn.: Appleton-Century-Crofts, 1980), p. 976. Reprinted by permission of the publisher.

Table: Allocation of Interior Residential Water Use

Use	***Percentage***
Drinking and cooking	5
Bathing	30
Toilet flushing	40
Laundry	15
Dishwashing	5
Miscellaneous	5
Total	100

Source: J. M. Last, in Maxcy-Rosenau, eds. *Public Health and Preventive Medicine*,11th ed. Reprinted by permission of the publisher.

400 feet since 1930 because of overpumping (Texas Department of Water Resources, 1984b). In the Second National Water Assessment, the U.S. Water Resources Council (1978a) lists actions that could be taken before pumping becomes uneconomical in the areas with declining water levels These include locating alternate water sources, developing artificial recharge of the aquifers, relocation of water-requiring activities, and reduction of water use through better water management. Groundwater overdraft is considered to be serious when it continues indefinitely, causing the exhaustion of the resource, associated economic and social dislocations, and perhaps general

weakening of a region's economy. Such potential problems are developing in California, the High Plains regions, Florida, New Mexico, Colorado, and Arizona, but many states are successfully dealing with the problem (U.S. GAO, 1980). Although this book deals with changes in groundwater quality, changes in quantity through overdraft are briefly mentioned here, as they can affect the quality of the remaining groundwater in the region.

Profile of an Aquifer

In order to provide a picture of the importance of groundwater, it is useful to describe a specific aquifer, its characteristics and uses, and the problems that affect it as a natural resource. A study by the High Plains Study Council provides such data for the Ogallala Aquifer. The Ogallala is an exceptional aquifer in both extent and volume, running as it does through parts of eight states. It is an integral part of the burgeoning economy of those states, as well as of the environment. Because of its unusual size and certain other characteristics, however, it should not be viewed as a representative aquifer.

The Ogallala Aquifer is a major source of groundwater in the High Plains region, underlying a land area of approximately 33 134,000 square miles, three times the size of New York State, or about the size of California. The formation extends from southern South Dakota to northwestern Texas and transects portions of six other states: Nebraska, Wyoming, Colorado, Kansas, Oklahoma, and New Mexico. It contains an estimated quadrillion gallons, or 3.25 billion acre-feet of water, the equivalent of Lake Huron. The aquifer fuels a $30 billion a year agricultural economy; irrigates over 12 million acres of farmland, or 20% of the irrigated acreage in the United States; and helps support a population of nearly half a million people. Furthermore, 40% of the beef consumed in this country is fattened in the aquifer area. Such heavy exploitation of the vast groundwater resources in the Ogallala has resulted in serious depletion in some areas. These problems are now the subject of intensive study.

The Ogallala was formed during the early Pleistocene time, some two million years ago, from glacial outwash of the ancestral Rocky Mountains that consisted of gravel, sand, and finer debris that was caught up by streams of meltwater running away from the glacier. The outwash settled unevenly throughout the High Plains, giving the aquifer water-storage capacities that varied according to the depth and content of the outwash. In Nebraska, where two-thirds of the Ogallala's waters lie and where the aquifer is 1,000-1,500 feet thick, the aquifer has a considerably greater storage capacity per unit of area than it does in the South Plains of Texas, where it is less than 100 feet thick. At present, there are an estimated 2 billion acre-feet of water in storage in Nebraska compared with 350 million acre-feet in Texas and 100 million acre-feet in Oklahoma.

Annual withdrawals from the entire aquifer average 5 or 6 million acre-feet (U.S. Water Resources Council, 1980). The Ogallala's extensive groundwater resources were virtually unknown until the early part of this century. The first wells were drilled into the formation over 90 years ago, but the land was used primarily for cattle grazing and dryland wheat production well into the 1900s. The start of irrigated farming, with the use of high-capacity pumps, 34 after World War II brought rapid economic growth but also caused the water table of the aquifer to decline steadily, in some areas by as much as 100 feet, and even to the point of drying up.

Gaines County, Texas, has dried up its groundwater resources entirely. The aquifer is very slow to recharge. Because of its high rate of evaporation and its high percentage of impermeable soils, the water is not replenished at the rate at which it is used. The depth of the water table nearly everywhere in the Ogallala formation is 50 feet or more below the roots of plants. Plants get the first opportunity at capturing any infiltrating rainfall, making that portion unavailable for pumping. Overall, the amount of water being overdrawn annually from the aquifer exceeds 3 million acrefeet. Annual recharge from precipitation averaging 12 to 22 inches per year, irrigation return-flows, and some stream-bed percolation is estimated to be 0.27 million acre-feet. The state of Texas has estimated that, at the current rate of use and without an imported water supply, approximately 40% of the now-irrigated acreage in the High Plains of West Texas will have to revert to dryland farming or be abandoned by the year 2000, and 60% will have to be abandoned by the year 2020. The depletion of the Ogallala began to attract public attention in the early 1970s. Members of Congress representing the Great Plains states of Colorado, Kansas, Nebraska, New Mexico, Oklahoma, and Texas encouraged the development of legislation that would mandate intensive study of the problem. A $6 million study was authorised by public law in 1976, and the High Plains Study Council was formed under contract with Camp, Dresser & McKee. The Council approved a study design with the following objectives in 1977:

1. to determine the potential development alternatives for the High Plains,

2. to identify and describe policies and actions required to carry out promising development strategies, and
3. to evaluate the local, state, and national implications of these alternative development strategies.

It is a comprehensive resource and economic-development study of the area served by the Ogallala. Major emphasis is being given to improving the water supply by local conservation and by improved practices of irrigation and agricultural management and to interbasin transfers of surplus water from adjacent basins, the Missouri River, and streams in Arkansas. Effective, comprehensive, interstate management of the 35 Ogallala groundwater resource is a further major element of the study.

Currently there is little or no public support for comprehensive, interstate management of the aquifer, although interdependencies are widely recognised. Laws in the states vary widely. Oklahoma and Texas are the only states of the High Plains with private ownership of groundwater. The groundwater in the six other states is dedicated to the people of the state, and allocation is administered by state and local officials. In all eight states, well registration or permitting systems exist.

Contamination of the Ogallala aquifer does not, to date, appear to be a major problem or concern. There has been some contamination around major population centres like Lubbock, Texas, but the depletion problem is of more immediate concern. A study by Camp, Dresser & McKee predicts that 5.1 million acres of irrigated land in six Great Plains states will dry up by the year 2020. A continuation of present trends will mean the loss of 1.6 million irrigated acres in Kansas, 1.2 million in Texas, 260,000 in Colorado, 224,000 in New Mexico, and 330,000 in Oklahoma. Twenty recommendations have been made for remedial action, but only one is an actual cure. The Army Corps of Engineers has proposed a system of huge canals that would import water from South Dakota, Missouri, and Arkansas. The cost of the project (from $3.6 billion to $22.6 billion) is prohibitive, however, and stopgap efforts like conversion to dryland farming will probably be the most immediate solutions to the depletion problem.

Uses of the Ogallala Groundwater Resources In a Few Selected States

In New Mexico the aquifer is heavily developed from southern Quay County to southern Lea County. The thickest and most productive part of the aquifer is in northern Lea County. Individual well yields in this area can be as much as 1,600 gallons per minute. In southern Lea County, where the aquifer is thinner, wells yield from 300 to 1,000 gpm, but water levels have been declining at a rate of almost 3 feet per year since 1950. In Oklahoma the Ogallala is the most important aquifer, underlying most of the panhandle. Deposits of sand, gravel, and minor amounts of clay store high-quality 36 groundwater. The aquifer is as much as several hundred feet thick, contains more than 100 million acre-feet of available water, and supplies most of the water requirements of the panhandle. It is used to irrigate 135,000 acres, to supply water for the industrial needs of the Keyes helium plant and the natural gas industry in the panhandle, and to supply all the public and domestic needs in the area.

In Texas the aquifer formation consists of interfingered and intergraded lenses and layers of sand, gravel, silt, clay and caliche, a crust of calcium carbonate. The High Plains is divided by the Canadian River into the northern High Plains and the southern High Plains. The northern area comprises approximately 9,300 square miles. The zone of saturation in most places is 100 to 500 feet thick. Irrigation has developed more slowly here than in the southern High Plains, which is the area of greatest groundwater development in Texas. This area includes about 25,000 square miles and has the most serious depletion problems of the entire aquifer. According to a 1985 USGS report, groundwater withdrawals in the southern High Plains have been averaging 6,500 million gallons a day (mgd) (USGS, 1985). The rate of recharge has been 125 to 134 mgd.

The Texas Department of Water Resources has made estimates and projections of groundwater availability in the Ogallala through the year 2030. It has determined the annual effective recharge to be 143 billion gallons. The amount of water in recoverable storage as of 1980 was determined to be 125,609.27 billion gallons. (Recoverable storage is defined as that portion of the underground reservoir capacity that, it is estimated, is physically capable of yielding water economically.) The projected annual average of groundwater availability was then determined through 2030. (Annual average of groundwater availability is defined as the estimated sustainable annual yield, or effective recharge, plus that amount of water that can be recovered from storage over a specified period of time without causing irreversible harm, such as land-surface subsidence or water-quality deterioration). The projections are as follows:

1990	2,132.17 billion gallons
2000	2,678.33 billion gallons
2010	2,495.95 billion gallons
2020	1,959.99 billion gallons
2030	1,491.09 billion gallons

The remaining recoverable storage in 2031 is projected to be 49,696.48 billion gallons (Texas Department of Water Resources, 1984c).

Groundwater Contamination:

- Groundwater contamination incidents have been reported from all parts of the United States.
- More contamination is likely to be discovered in the future, especially if a comprehensive national survey is undertaken.
- There are many and varied sources of contamination: natural and anthropogenic, point and nonpoint, planned and inadvertent.
- Types of contaminants found in groundwater range from simple inorganic ions, such as chloride, nitrate, and heavy metals, to complex synthetic organic chemicals.
- The types of reported contamination problems, and their sources, vary from one region of the country to another and are influenced by climate, population density, intensity of industrial and agricultural activities, and the hydrogeology of the region.

The mobility of our society, as well as the distribution of industry and agriculture, depends upon an available supply of clean water. Nevertheless, instances of groundwater contamination have been found in most sections of the country. For the purpose of this book groundwater contamination will be defined as the addition to water of elements, compounds, or pathogens that alter its composition.

The pollution of groundwater occurs when the concentrations of the contaminants render the water unfit for present and future uses. One of the major difficulties with groundwater contamination is that it occurs underground, out of sight. The sources of pollution are not easily observed, nor are the effect of pollution often seen until irreversible damage has occurred. There are no obvious warning signals such as fish kills, discoloration, or stench that are often early indicators of surface water pollution. Where contamination affects pumping wells, such indicators may occur, although many commonly found contaminants are both colourless and odourless, which makes detection difficult. They occur in concentrations that, if ingested, may cause long-term chronic illness, rather than acute poisoning. Many chemicals now found in groundwater have not been tested for their effects on human health. The tangible effects of groundwater contamination usually come to light long after the incident causing the contamination has occurred. The long time lag between occurrence and detection is a major problem.

Groundwater can be contaminated by a variety of compounds, both natural and man-made. Contamination due to man has occurred for centuries, but industrialisation, urbanisation, and increased population have greatly aggravated the problem in some areas. Miller (1981a), Braids (1981a) and Saar (1985) summarised the mechanisms of contamination and attenuation. Freeze and Cherry (1979) give a detailed account of these mechanisms, as well as of the movement of a contaminant in an aquifer.

Movement of Contaminants

A contaminant usually enters the groundwater system from the surface of the land, percolating down through the aerated soil and nonsaturated zone. The root zone may extend two or three feet into the soil. Many reductive and oxidative biological processes take place in this zone that may degrade or biologically change the contaminants. Plant uptake can remove certain heavy metals; microbial fixation and other biological processes can also remove a fraction of the contaminants, the size of the fraction being dependent on the nature of the contaminant. For example, depending upon conditions, iron and manganese solubility can be affected by microbial oxidation and reduction in the soils. In the deeper geologic material that consists mainly of humus and weathered rocks, there is a lessening of such biological processes. Attenuation of contaminants in this zone may occur by surface adsorption as ions in the contaminants are attracted to the charges on the clay particles. Other contaminants may be removed by insoluble organic matter complexing, giving rise to complexed humic acids. Microbial action may influence redox potentials and cause the release of inorganic ions during decomposition. The susceptibility of different contaminants to differential attenuation varies. Present knowledge indicates that more different attenuation processes are active in the nonsaturated zone than in the aquifer; and that the amounts of degradation are larger. However, many significant bio-conversions are being discovered in anaerobic subsurface environments. In fact many compounds that persist under aerobic conditions are metabolised in anaerobic conditions.

In aquifers composed of unconsolidated granular media, contaminant movement is determined by groundwater flow rates and flow paths which are governed by advection, diffusion, and dispersion, by the density of the contaminant and biological, chemical and physical interactions with the material through which it flows. Cherry (1984) states that "Contaminant migration in fractured media is determined by advection and dispersion in the fractures and by diffusive mass transfer between the flowing water in the fractures and the stagnant water in the porous but low permeable matrix."

Specific contaminants move in different ways. Liquid petroleum hydrocarbons, for example, would be expected to move downward through the aquifer; however, they may spread out on the top of the aquifer if they are present in sufficient quantities. They usually do not move very far from the point of entry because of the limiting effect of surface tension between the liquid and the water-coated grain of the unconsolidated material. However, dissolved chemicals from these petroleum hydrocarbons move with the groundwater.

Industrial liquids that are halogenated hydrocarbons have a viscosity less and a density much greater than those of water. The fractions that dissolve in water move with the groundwater. However, they may move downward, settle out in the aquitard, and (depending on the slope) may move downslope and not in the direction of the bulk flow of the groundwater.

The migration of contaminants through unconsolidated granular media is better understood than is their migration through fractured material. Unfortunately, contaminant migration through subsurface materials is a highly complex and poorly understood science. Much of the quantitative data obtained for the prediction of contaminant migration from theoretical models and laboratory data is of questionable use when applied to the field. Problems arise mainly from variable geologic conditions (heterogeneous, anisotropic, and fractured subsurface materials), and from the many types and characteristics of the contaminants themselves. Prediction of plume movement and behaviour is made difficult because geologic formations are not often uniform. Layered beds and lenses may cause fingering or separation of the plume. Different geologic materials may retard or enhance the movement of the plume. The nature of the plume is further affected by the reactivity of the constituents of the contaminant. Concentrations can be altered by a variety of chemical and biochemical reactions, including adsorption-desorption reactions, acid-base reactions, solution-precipitation reactions, oxidation reduction reactions, ion pairing, and microbial cell synthesis. The plume continues to move with groundwater flow unless it is blocked and eventually reaches points of groundwater discharge such as streams, wetlands, lakes, and tidal waters. Because very little dilution takes place, concentrations of contaminants are often much higher in groundwater than in surface water.

Sources of Contamination

Sources of groundwater contamination may be divided into three main categories:

1. natural pollution
2. waste-disposal practices
3. nondisposal sources due to human activities.

The degree of threat posed by these sources of contamination depends upon the concentration of the contaminant, its toxicity at that concentration, the volume of groundwater affected, the hydrogeological conditions, the uses made of the water from that particular aquifer, the population affected by such uses, and the availability of an alternate water supply.

Groundwater-pollution problems and their sources have been the object of numerous summaries. An incomplete but illustrative listing includes Keeley, 1976; Fuhriman and Barton, 1971; van der van der et al., 1975; Miller et al., 1977; Scalf et al., 1973; Miller et al., 1974; U.S. EPA, 1980a, 1978a, 1978b; U.S. GAO, 1978, 1980; U.S. Water Resources Council, 1978a, 1978b, 1978c. A comprehensive survey of sources of contamination due to wastedisposal practices was completed for the 1977 Report to Congress. However, the most comprehensive listing of sources thus far is that put together by the Office of Technology Assessment. OTA identified 33 sources of known groundwater contaminants and categorised them according to the nature of their release.

Changes in the Composition of Groundwater Due to Natural Processes

All groundwater contains some dissolved salts. Mineralisation of groundwater due to leaching is a significant source of saline groundwater in the arid areas of the country and is greatest in the areas of lowest precipitation. In areas where the water table is near the surface, evaporation and transpiration further concentrate the salts in the remaining water. Natural leaching has been a significant source of contamination in the arid southwest and south-central area comprising Arizona, California, Nevada, Utah, Arkansas, Louisiana, New Mexico, Oklahoma, and Texas. In this area, natural leaching has exceeded in importance anthropogenic sources of groundwater contamination. Many of the inorganic substances found in groundwater are largely the result of natural contamination, although human sources contribute significant amounts as well. Chlorides are a widespread contaminant. Sulfates, nitrates, fluorides, and iron are also common natural contaminants that occur in localised natural deposits, often causing the groundwater in those areas to exceed EPA standards. Radioactivity from uranium deposits has caused problems in Texas, Illinois, Iowa, Oklahoma, Nevada, Arkansas, and New Mexico. Arsenic may be a local problem in the thermal springs of the northwest region (van der Leeden et al., 1975). Other examples of such localised problems occur throughout the country. Water from fault zones or of volcanic origin can contain high levels of salts or toxic chemicals. As mentioned in the section characterising groundwater, the aquitards that confine the aquifers are not totally impermeable. A change in the composition, thickness, or continuity of the aquitard may permit leakage from one aquifer to another. The leaking of a saline aquifer into a potable aquifer may cause deterioration of the quality of water in the potable aquifer. Although natural alteration of the chemical composition of groundwater cannot be prevented, it may be fit for some uses with or without pretreatment.

Contamination Due to Waste-Disposal Practices

The 1984 OTA report to Congress *Protecting the Natio's Groundwater from Contamination* had the best estimate of the number of sources of groundwater contamination from waste disposal at the time that this report was compiled. The major sources described in this report and elsewhere will be summarised below. It should be noted that many of the numbers given are estimates and that definitive data on waste-disposal practices often were not, and are not, readily available.

Waste from natural and manufactured products is often stored on or beneath the land surface. In fact, burying wastes seems an ingrained cultural phenomenon. Such burial of wastes is now regulated by the statutes. Sources of contamination are derived from all aspects of our lives, including industry, agriculture, and government. One estimate is that between 50,000 and 75,000 chemicals are now being used and distributed through the environment and that an additional 700 to 800 are being added each year.

Table: *Inorganic Substances Found in Groundwater*

Substance	***Concentration (mg/l)***
Aluminum	0.1-1,200
Ammonia	1.0-900
Antimony	—
Arsenic	0.01-2,100
Barium	2.8-3.8
Beryllium	less than 0.01
Boron	—
Cadmium	0.01-180
Calcium	0.5-225
Chlorides	1.0-49,500
Chromium	0.06-2,740
Cobalt	0.01-0.18
Copper	0.01-2.8
Cyanides	1.05-14
Fluorides	0.12-250
Iron	0.04-6,200
Lead	0.01-5.6
Lithium	—
Magnesium	0.2-70
Manganese	0.1-110
Mercury	0.003-0.01
Molybdenum	0.4-40
Nickel	0,05-0.5
Nitrates	1.42-433
Nitrites	—
Palladium	—
Potassium	0.5-2.4
Phosphates	0.4-33
Selenium	0.6-20
Silver	9.0-330
Sodium	3.1-211
Sulfates	0.2-32,318
Sulfites	—
Thallium	—
Titanium	—
Vanadium	243.0
Zinc	0.1-240

— indicates that the substance has been detected in groundwater but no concentrations have been reported. (Source: OTA, 1984.)

In 1984, the OTA listed more than 200 substances that have been detected in groundwater; these include 175 organic chemicals and 50 inorganic chemicals, organisms, and radionuclides. For disposal, however, most are reacted with other chemicals, immobilised, or otherwise treated and burned.

Individual Sewage Disposal Systems

The 1980 census estimated that there were approximately 22 million septic systems serving nearly one-third of the nation's population. Annual flow to an individual septic tank ranges from approximately 49,000 to 75,000 gallons per household. The total annual flow to all domestic systems can be estimated to be from 1.08 trillion to 1.65 trillion gallons, making these systems the highest-ranking source of wastewater discharged directly into groundwater. The three methods of onsite domestic waste disposal are septic tanks, with their subsurface disposal system; the less satisfactory cesspool, commonly found in older installations where there is a deep layer of permeable soil; and finally the pit privy. Properly constructed septic-tank systems and privies permit effective treatment of human waste. Cesspools work only in coarse or highly fissured materials and have a high potential for contamination. The septic tank has several important functions. It separates solid and liquid wastes, stores solids and floatable materials, and treats aerobically both stored solids and nonsettlable material. Soils with slow percolation rates, shallow soils, soils over permeable bedrock, and soils with permanent or periodic high water tables are not suitable for the use of conventional septic tank systems.

The leach beds for septic tanks typically consist of 500 to 600 square feet of drain bed, located 18 to 24 inches below the surface of the ground. The effluent is thus distributed over a wide area, and the shallow depth of the bed permits some evaporation and some uptake of contaminants by plants. Because it is near the surface, the drain field operates under aerobic conditions, with the result that good quality water reaches the groundwater. It is important that a zone of unsaturated soil occur between the leach bed and the water table so that the effluent from the septic tank is not discharged directly into the groundwater and hence into the aquifer. The degree of potential risk posed by these systems depends in large measure upon their design and installation and upon the hydrogeology of the area. When sited correctly they operate efficiently, renovating wastewater and returning good quality water to the watershed. When operating incorrectly, they can allow pollutants to enter the groundwater. This would be particularly 62 troublesome in areas where drinking water wells tap the same aquifer in the vicinity of the leach bed. The advantages of septic tanks include the following:

- Minimal maintenance is required.
- The cost of individual or community septic tanks is less than the cost of central wastewater collection and treatment plants.
- It is a low-technology system.
- The energy requirements are low when compared with centralised wastewater-treatment facilities.

The disadvantages include:

- The potential for groundwater contamination exists when the tanks are incorrectly sited, with regard to density of the individual units or soil characteristics or both.
- The systems require proper maintenance. Failure to provide this results in system overflows or pollution of nearby wells.
- Septic-tank cleaners may severely contaminate groundwater.

The constituents that pose the greatest threat to the quality of groundwater are nitrates, phosphates, heavy metals, inorganic ions (Na^+, Cl^-, $SO_4^=$, K^+, Ca^{++}, and Mg^{++}), and pathogenic organisms. In addition, toxic synthetic organic chemicals are becoming a more significant hazard due to the increased use over the last ten years of cleaning products containing such chemicals. It is difficult to show by mapping how regions are affected by sources of contamination. All contamination incidents are indicated by points on a map, but not all the groundwater in the area indicated is contaminated. Since the most important factors affecting the contamination of groundwater by sewage are the density of the individual units and the soil characteristics, the greatest potential for contamination is in the eastern region of the country. A septic-tank density of greater than 40 per square mile constitutes a region of potential contamination to groundwater.

Table: *Characteristics of Domestic Sewage*

Constituent	***Concentration *Typical Domestic Sewage***
* mg/l except for conductivity (micro-mhos/cm), and pH (pH units).	
Total Suspended Solids	200
Conductivity	700
Chemical Oxygen Demand (COD)	500
Biochemical Oxygen Demand (5 day BOD)	200

Contd...

Constituent	***Concentration *Typical Domestic Sewage***
Total Organic Carbon (TOC)	200
pH	8.0
Alkalinity (as CaCO)	100
Acidity (as CaCO)	20
Total Phosphorus	10
Total Nitrogen	40
Chloride	50
Calcium	50
Magnesium	30
Iron	0.1
Manganese	0.1

Source: U.S. EPA, 1973.

Land Disposal of Solid Wastes

According to the Resource Conservation and Recovery Act of 1976, solid wastes are defined as any garbage, refuse, or sludge from waste-treatment plants, water-supply treatment plants, or airpollution control facilities, and any other discarded materials, including solid, liquid, semi-solid, or contained gaseous materials, resulting from industrial, commercial, mining, and agricultural operations and from community activities. They do not include solid or dissolved material from domestic sewage, irrigation returnflows or industrial discharges that are point sources, or products and by-products of the nuclear industry. Solid wastes are considered hazardous when their quantity, concentration, or physical, chemical, or infectious characteristics cause, or significantly contribute to, an increase in mortality or an increase in serious illness, or pose present or potential hazards to human health or the environment when improperly treated, stored, transported, disposed of, or otherwise managed.

The solid portion of household wastes contains a high proportion of putrescible matter that is broken down by biodegradation, releasing carbon dioxide and methane gas (UNESCO, 1980). The leachate contains high concentrations of sulfate, chloride, and ammonia. Abundant quantities of cellulose from paper products may retard the movement of halogenated hydrocarbons by absorption.

***Figure** : Solid Waste Disposal Land*

Solid commercial wastes have a similar composition to household wastes. They may, however, contain greater quantities of 64 oils, phenols, and hydrocarbon solvents. Phenols are the most resistant to biological breakdown and may be leached. Domestic and commercial wastes are serious groundwater contaminants because of the numerous dissolved constituents and the high biological oxygen demand (BOD) of may of the constituents. The table does not include synthetic organic chemicals.

The solid components of industrial wastes vary with the source of production (UNESCO, 1980). Cyanide wastes are produced in metallurgical operations; sulfite wastes come from paper and pulp manufacturing; mercury is a waste product in the electrical industry; and the petrochemical industry produces several solid residues ranging from polychlorinated biphenyls (PCBs) to pesticides and herbicide residues to phenol-rich tar wastes. The amount of solid waste disposed of by various manufacturing industries is listed.

Other serious solid-waste contaminants can result from thermal power generation from the burning of coal, forming fly ash which, due to its high surface area volume ratio, is fairly reactive. Leaching of the deposited fly ash may initially give rise to high concentrations of arsenic, chromium, selenium, and chloride. Another contaminating product resulting from the burning of coal is the sludge formed by the aqueous scrubbing of flue gases. Sludges typically contain cyanide and heavy metals and are of low pH unless neutralised by the addition of lime. It has been found that mixtures of sludge, fly ash, and lime rapidly set into a low permeability solid which leaches less readily.

Table: *Summary of Leachate Characteristics Based on 20 Samples from Municipal Solid Wastes*

Components	***Median Value (ppm)***[a]	***Ranges of All Values (ppm)***[a]	
[a] Where applicable			
Alkalinity ($CaCO_3$)	3,050	0	-20,850
Biochemical Oxygen Demand(5-day BOD)	5,700	81	-33,360
Calcium (Ca)	438	60	- 7,200
Chemical Oxygen Demand (COD)	8,100	40	-89,520
Copper (Cu)	0.5	0	- 9.9
Chloride (Cl)	700	4.7	- 2,500
Hardness ($CaCO_3$)	2,750	0	-22,800
Iron, Total (Fe)	84	0	- 2,820
Lead (Pb)	0.75	<0.1	- 2.0
Magnesium (Mg)	230	17	-15,600
Manganese (Mn)	0.22	0	- 125
Nitrogen (NH_4)	218	0.06	- 1,106
Potassium (K)	371	28	- 3,770
Sodium (Na)	767	0	- 7,700
Sulfate (SO_4)	47	1	- 1,558
Total Dissolved Solids (TDS)	8,955	584	-44,900
Total Suspended Solids (TSS)	220	10	-26,500
Total Phosphate (PO_4)	10.1	0	- 130
Zinc (Zn)	3.5	0	- 370
pH	5.8	3.7	- 8.5

Source: U.S. EPA, 1977.

Contamination occurs when precipitation infiltrates and percolates through solid wastes at poorly designed land-disposal sites, forming a leachate of dissolved minerals, heavy metals, and organic chemicals. Leachate formed at disposal sites which are located in wetlands, flood plains, or where there is a shallow water table and that are constructed without natural or artificial barriers is likely to contaminate groundwater.

The types of land disposal sites for solid wastes include dumps, landfills, sanitary landfills, and secured landfills listed in increasing order of ability to protect the environment from the adverse effects of their use. Landfill regulations promulgated under the Resource Conservation and Recovery Act (RCRA) were issued in July 1982. The Comprehensive Environmental Response, Compensation and Liability Act (CERCLA) requires remedial actions at inactive sites that pose a threat to the environment and to human health. The U.S. EPA estimates that there are 93,000 landfills in the United States. Approximately

75,000 of these are onsite industrial landfills and 18,500 receive municipal wastes (U.S. EPA, 1985c). In 1983, another survey estimated that there are between 15,000 and 20,000 active landfills in the nation, including municipal landfills, open dumps, and nonhazardous waste facilities, operating, abandoned, or closed. (OTA, 1984).

Hazardous wastes may be disposed of in a landfill, dump, or impoundment. EPA requires that site locations and amounts and types of wastes stored be reported, under CERCLA by owners and operators of inactive sites, and under RCRA by present owners and operators of active privately owned industrial sites. About 199 active hazardous-waste landfill facilities (OTA, 1984) and approximately 20,000 abandoned hazardous-waste sites (EPA, 1985c) are known to exist across the country.

The amount of solid waste generated in this country has been estimated to be enough to cover 400 acres of land to a depth of 10 feet each day. It is estimated that in 1980, 3.74 pounds of solid waste were generated per person per day, amounting to 155.6 million tons of municipal wastes nationwide for the year. OTA (1984) also estimates that 30 million tons of solid wastes generated by utilities and 1 million tons of municipal sludge are disposed of in landfills per year. Between 40 million and 140 million wet tons of nonhazardous industrial solid wastes are disposed of annually (OTA, 1984). In 1981 hazardous-waste facilities received 14.7 billion gallons of both liquid and solid hazardous wastes (CEQ, 1984).

Collection, Treatment, and Disposal of Municipal Wastewater

The collection, treatment, and disposal of municipal wastewater may be a problem in urban areas. Municipal wastewaters include domestic wastewater, storm water and its associated debris, and industrial wastes. They are collected by sanitary sewer systems and transported to treatment sites. Approximately 170 million Americans were served by sewer systems in 1984.

Figure : *Using treated municipal wastewater can be quite difficult*

Leaks may occur in the system due to design, age, disruption by tree roots, seismic activity, or poor construction. The lagoons and ponds used in wastewater treatment operate under anaerobic, aerobic, or facultative conditions.

The total volume of sewage handled by one of these plants is estimated to be approximately 0.667 mgd. Lagoons are often constructed without adequate seals, thus promoting leakage under certain geologic conditions and posing a threat to groundwater.

Another possible route of groundwater contamination is by the land disposal of treated wastewater, which could alter the chemical constituents of the natural groundwater. Land disposal methods include agricultural irrigation, rapid infiltration ponds, overland runoff, and discharge into dry stream beds and ditches. Methods involving spraying onto land are not suitable for use in freezing winter months.

Industrial and Other Wastewater Impoundments

Under the Safe Drinking Water Act of 1974, EPA undertook a national survey of surface impoundments, which was carried out from 1976 to 1978, and the interim report was published in 1978. The final report, released in 1983, contains data from 1978 through 1980. These reports define surface impoundments as depressions in the land (pits, ponds, lagoons, and pools) containing liquid, semi-solid, and solid wastes. In the final report EPA stated that the number of sites located in the United States totalled approximately 80,263 and contained more than 180,000 impoundments.

Thirty-one percent of the sites were oil 68 and gas brine pits, 25% municipal sites, 18% agricultural sites, 15% industrial sites, and 9% mining sites. At the time of the survey it was estimated that approximately 70% of the industrial sites, 84% of the agricultural sites, and 78% of the municipal sites remained unlined. More than 98% of the sites were located in areas with proximity to potential water supplies. Approximately 15% of the sites (excluding oil- and gas-related facilities) contained hazardous wastes. New Mexico was estimated to have the largest number of sites, with 12,901, and Rhode Island the fewest, with 44. The regulations under RCRA may have affected the validity of these estimates, but no recent figures are available.

Table: *Summary Statistics for Located Active Surface Impoundment Sites*

Category	***Located Sites****		***Assessed Sites****		***Located Impoundments****	
Industrial	11,760	(15%)	8,662	(28%)	27,912	(15%)
Municipal	19,746	(25%)	10,822	(34%)	37,185	(21%)
Agricultural	14,850	(18%)	6,646	(21%)	19,437	(11%)
Mining	7,364	(9%)	1,552	(5%)	25,038	(14%)
Oil & Gas Brine Pits	24,990*	(31%)	3,354	(11%)	65,488	(36%)
Other	1,553	(2%)	350	(1%)	5,913	(3%)
Total	80,263		31,386		180,973	

* SIA site numbers for the mining and oil & gas brine pit sites are usually related to lease or field data, not to actual ownership, and should not be referred to as the actual number of legal sites. The number of located impoundments would be a closer approximation for these two categories. *Located sites:* Total number of facilities identified in the inventory. *Assessed sites:* Total number of facilities evaluated in the inventory. *Located impoundments:* Total number of impoundments identified in the inventory. The number is larger than located sites since many facilities had more than one impoundment.

Source: U.S. EPA, 1983.

Table: *Estimated Number of Located Impoundments and Impoundment Sites for All Categories, by State*

State	***Active Sites***	***Impoundments***	***Abandoned Sites***	***Impoundments***
Alabama	1,063	1,696		
Alaska	129	179	27	31
Arizona	641	1,615	20	48

Contd...

State	***Active Sites***		***Abandoned***	
	Impoundments		***Sites Impoundments***	
Arkansas	861	6,806	12	24
California	1,750	7,577	31	101
Colorado	1,267	3,631		
Connecticut	305	758	83	188
Delaware	52	131		
Florida	3,644	5,610		
Georgia	1,363	1,640	65	70
Hawaii	94	296		
Idaho	253	538		
Illinois	4,471	6,677	591	745
Indiana	1,897	2,678		
Iowa	1,784	2,539		
Kansas	1,970	3,328		
Kentucky	640	922		
Louisiana	984	2,804	49	77
Maine	132	319	41	133
Maryland	443	836	18	25
Massachusetts	341	2,211	38	208
Michigan	2,187	3,505	76	135
Minnesota	1,952	2,733	135	160
Mississippi	1,306	1,667	52	58
Missouri	3,657	4,683	507	507
Montana	668	1,266		
Nebraska	1,632	3,254		
Nevada	216	541	10	382
New Hampshire	152	274	2	4
New Jersey	311	896	46	129
New Mexico	12,901	17,746	39	116
New York	1,070	1,837	40	52
North Carolina	709	1,051	26	30
North Dakota	667	1,168		
Ohio	3,287	16,537		
Oklahoma	2,935	4,538		
Oregon	370	714		
Pennsylvania	7,346	34,224	234	627
Rhode Island	44	133	3	13
South Carolina	1,868	2,259		
South Dakota	606	926		

***Table:** Estimated Number of Located Impoundments and Impoundment Sites for All Categories, by State* (continued)

State	***Active Sites***	***Impoundments***	***Abandoned Sites***	***Impoundments***
Tennessee	746	1,325	59	151
Texas	3,672	10,740	170	365
Utah	211	2,345		
Vermont	214	274	10	17
Virginia	1,212	1,975	30	62
Washington	403	1,049		
West Virginia	688	1,861	66	186
Wisconsin	1,022	1,720	49	82
Wyoming	1,146	1,738		
Total	79,818	180,328	2,519	4,726

Source: U.S. EPA, 1983.

Impoundments which are used for the treatment, storage, or disposal of wastes are natural or man-made, and may or may not be lined. Discharging impoundments are designed to discharge regularly into bays, oceans, lakes, or streams. The liquid in nondischarging impoundments is lost by evaporation or by seepage into the soil. The types of wastes received by these sites range from sewage wastes, industrial wastes including those from air scrubbers, cooling tower blow down, and ash residues and oiland gas-extraction wastes to animal feedlot and other agricultural wastes. Of these, the largest users of surface impoundments are the oil and gas industries. Evaporation ponds were used extensively in the South for oil-brine disposal. Evaporation ponds usually lose more fluid by infiltration than by evaporation, especially in the more humid regions of the country. EPA promulgated regulations for hazardous waste surface impoundments in July 1982. In addition, state regulations exist in most oil-producing states.

The chemical character of the waste varies enormously and may include suspended and dissolved solids, pathogenic organisms, oil and grease, detergents, heavy metals, and toxic organic chemicals. It may have a high biological oxygen demand (BOD), chemical oxygen demand (COD), and total organic carbon (TOC). The pH also varies. Not all surface impoundments contain hazardous wastes. Some are used as holding ponds prior to the treatment of the fluids. Because the chemical character of the wastes varies, only a minimal number of impoundments would hold all of the contaminants listed. The type of groundwater contamination known to occur as a result of leaking impoundments reflects the nature of the fluids in the impoundments. Dissolved materials, both inorganic and organic, move into groundwater by direct seepage, while solids may be leached by precipitation or waste fluids. As with all sources of contamination, the sorptive capacity and low permeability of some soils may slow or impede the movement of some of the contaminants.

Land Spreading of Sludge

When the wastewater from municipal and industrial wastes is treated, a residue of sludge remains. Beginning in November 1981, all facilities disposing of sludge considered hazardous on land were required to have a groundwater-monitoring system in place under Resource Conservation and Recovery Act regulations. In 1982 there were at least 2,463 publicly owned treatment facilities applying liquid and thickened sludge on land surfaces, and an additional 485 facilities in operation or under construction using sludge spray irrigation practices (OTA, 1984). Of the 6.8 million dry tons of sludge produced by municipalities in 1982, 24%-29% was spread directly onto crops (OTA 1984). The purpose of land treatments is to biodegrade the organics and to immobilise the inorganics. The potential for contamination exists, however, since the chemical constituents and viruses in sludge may be leached by precipitation after the sludge is spread on the land. Sludge from manufacturing industries may contain either useful agricultural chemicals or toxic compounds, and it varies from degradable to very refractory. Sludge is stabilised and may be dewatered before land application. The higher the proportion of clay and organic colloids in the soil, the greater the capacity for the heavy-metal ions present in sludge to be immobilised. The main hazardous constituents of the sludges include heavy metals and organic chemicals, such as dyes, inks, oils, pesticides, detergents, organic solvents, and polynuclear aromatic hydrocarbons (PAHs); organic chemicals are typically trace contaminants. Sludges from the municipal treatment of wastewater or the biotreatment of industrial wastewater include biological components such as bacteria, viruses, fungi, algae, protozoans, rotifers, and other parasites such as worms and flukes. The constituents that may be found in various types of industrial wastes are enumerated in the 1977 Report to Congress. As of 1985, EPA reported seven states that had experienced groundwater contamination due to land spreading of sludges.

Land farming of sludge has been used for nearly 30 years by the petroleum industry. From its research on this method of disposing of oil sludge, the Solid Waste Management Committee of the American Petroleum Institute concludes that if the area farmed is carefully matched to the amount of organic material that must be degraded, the type of land farming practiced, and the degradation rate of the sludge, then the underlying soil and groundwater will not be significantly contaminated.

Brine Disposal Associated with the Petroleum Industry

Groundwater contamination associated with oil production has been documented in at least 21 states (OTA, 1984). During 1983 the total production of crude petroleum was more than 3 billion barrels, valued at more than $83 billion. The principal oil-producing states in 1983 were Texas, Arkansas, Louisiana, and California. Brief mention of the disposal of brines by the use of surface impoundments was made above. Oil production is usually accompanied by the production of saline wastewater in amounts that vary with production procedures. One estimate is that an old well may produce 100 barrels of brine for each barrel of oil. Others estimate that the production of 8 million barrels of crude oil produces 30 million barrels of saltwater and that the ratio of brine to crude oil recovered is 10:1. Still others estimate a ratio of 4:1 brine to oil (OTA, 1984).

In the early days of oil and gas production, the brine was often disposed of in unlined pits. The extensive use of such "evaporation" pits that leaked their contents has caused groundwater contamination in the oil-producing regions. Use of unlined pits has been completely eliminated in some states and partially eliminated in others. For example, unlined pits were banned in Texas in 1969. Other states permit the use of small pits in remote areas if they pose no threat to groundwater.

Present practices for brine disposal include injection of the brine into deep underground formations that are deemed unsuitable for other purposes and reinjection of brines into oil-producing formations to enhance oil recovery. These injection practices, if not properly controlled, pose a potential threat to groundwater.

Past oil production practices, including well drilling, completion, and abandonment practices, also were not as strictly controlled as they are now. Some wells were improperly equipped, and abandoned wells were improperly plugged, thereby providing a potential for groundwater contamination. In 1983 abandoned production wells numbered around 1.2 million (OTA, 1984). The need to protect groundwater was recognised, however, and state oil and gas regulatory agencies now have regulations concerning brine injection, well completion, and plugging practices. The Interstate Oil Commission sponsored a study in 1978 that concluded that in Texas, Arkansas, Louisiana, New Mexico, and Oklahoma, reinjection of brines had resulted in fewer instances of groundwater contamination than had the use of evaporation pits. It therefore believes that the practice does not pose a major threat to the groundwater supplies.

Most oilfield brines are corrosive, but the corrosion rate of old and new well casings is unpredictable, and frequent monitoring and surveillance is therefore carried out. Although brine is the main threat to the quality of groundwater related to oil production, minor threats are also posed by the oil and gas themselves, by drilling fluids, by chemicals used in treating wells, by corrosion inhibitors, and by other chemicals. Oil drilling may entail drilling through freshwater aquifers, thus raising the potential for contamination.

Disposal of Mine Wastes

Mining creates conditions and products conducive to groundwater contamination. The recharge and movement of groundwater may be affected by the mining process. Domestic mining for minerals other than organic fuels was more than a $20 billion industry in 1983. Domestic mining for minerals other than organic fuels was a $10 billion industry in 1974. Coal has enjoyed a comeback as an energy source due to the goal of increased self-sufficiency in energy, and it is estimated that between one-third and one-half of the world's total reserves are found in the United States. Techniques of surface and underground mining pose different threats to groundwater quality, although the threat from waste disposal from the two is similar. Although such mining changes the environment, good reclamation may stabilise some of these changes without causing serious damage to the groundwater. Sand, clay, gravel, and stone quarrying are not considered a serious threat to groundwater quality. Threats from the remaining categories of mines include those posed by leachate from tailings piles and drainage from strip, surface, and underground mines. Dewatering of mines, involving a regional lowering of the water table, may have three results: a decrease in the amount of groundwater supplied, a lowering of the water level below the intakes of productive wells, and the oxidation of exposed minerals. Coal is often found in conjunction with iron pyrites (FeS_2) and other sulfides, and the oxidation of these sulfides and the presence of percolating waters produce sulfuric acid and an increase in Fe^{++} and $SO_4^{=}$, leading to acidic groundwater, a serious problem in the coal-mining areas of Pennsylvania and Appalachia. It is very costly either to neutralise acid mine drainage or to backfill abandoned mines in order to prevent the problem. Sealing the mines with airor watertight seals has not proved very successful. A report by the National Academy of Sciences summarizes these and other effects that coal mining may have on groundwater quality.

Metal mining produces drainage waters that have lower concentrations of iron and sulfate than do colliery waters, but that have higher concentrations of dissolved heavy metals due to the lower pH caused by the oxidation of metal sulfides. Wastewater from metal mining may also contain organic flocculants from the onsite processes of screening and dressing the ores. The wastes from such operations, which can be both toxic and radioactive, may contaminate groundwater. The tailings ponds used in the waste-disposal operations of mines may contribute contaminated leachate.

The waste products of quarrying for stone, lime, gypsum, and the like are generally inert, and their only contributions to the water that percolates through them is the possibility of an occasional increase in suspended-solids if crushed material is present. Quarrying does, however, increase the potential for local groundwater contamination in other ways. The removal of soil reduces the possibility of an attenuation of any spills or leakages that may occur in the area and increases the potential for contamination, especially if the bedrock is fractured.

Because of the serious nature of surface water contamination by mining activities, more attention has been given to it than to groundwater contamination, although aquifers in regions with a long history of mining have been written off as sources of water supply. Since then coal mining on a large scale has been initiated in the West, particularly in North Dakota, Wyoming, and Montana. The EPA (1977) estimated that 3.6 million tons of acid may be generated annually from the approximately 200,000 acres used across the nation for disposal of coal-mining wastes. The EPA (1977) also estimated that 10% of the acid enters groundwater each year. Mining operations generate approximately 2.3 billion tons of total waste material annually, including radioactive and nonradioactive waste piles and tailings. The most serious threat to groundwater is posed by the metal mines, particularly uranium and copper mines, which produce waste containing dissolved toxic materials (e.g., arsenic, sulfuric acid, copper, selenium, and molybdenum), as well as radioactive materials.

Deep-Well Disposal of Liquid Wastes

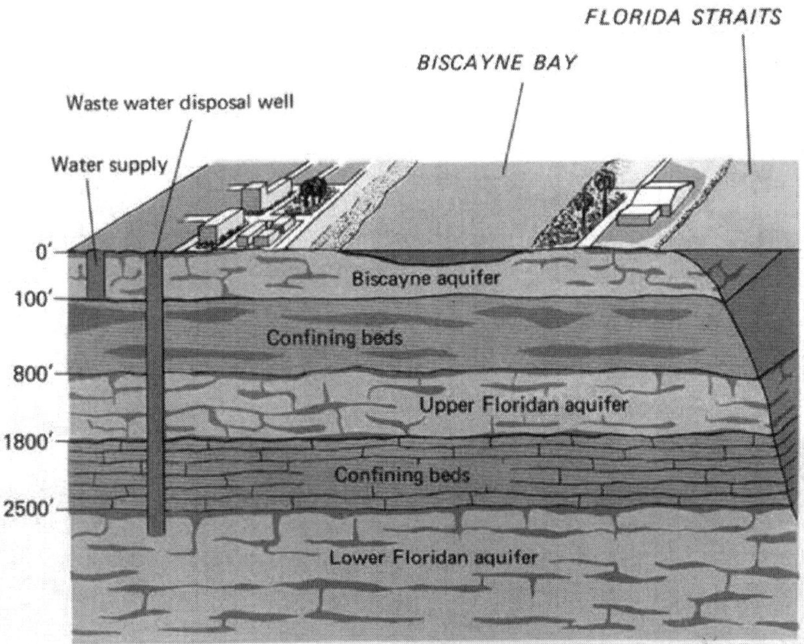

Figure : *Some sewage effluent is disposed of by deep-well injection*

The injection of a variety of liquid wastes into deep wells to avoid having them contaminate the biosphere was becoming a widely adopted waste-disposal practice in the late 1970s, mainly because there were stringent regulations governing disposal into surface waters. Such wells are less widely accepted now in certain states. It has been estimated that there are more than 221,000 wells in the United States injecting liquid wastes underground: approximately 280 are used for hazardous wastes; 140,000 are used either for brine disposal or for the injection of fluids used in enhanced oil-recovery process; 40,000 are used for agricultural drainage, urban runoff and sewage disposal; and at least 12,000 are in operation for solution mining (OTA, 1984).

Because of the hazardous nature of the injected fluid often radioactive, toxic chemical, petrochemical, and pharmaceutical wastes that are difficult to treat deep-well disposal may pose a great threat if improperly practiced. Eighteen states have already experienced contamination through injection wells. The Underground Injection Control Program of the Safe Drinking Water Act, which prohibits direct injection into drinking water aquifers, and regulations issued under the Resource Conservation and Recovery Act provide such controls.

All injection wells are in the depth range of 660 to 13,200 feet, most being between 990 to 6,600 feet. Injection zones are usually located in sandstone, basalt, and carbonate rocks. Injection occurs under pressure, and injection rates vary from 500 to 370 gallons per minute. Injection causes hydrodynamic changes in the aquifer, including the formation of a mound in the potentiometric surface extending in the direction of the regional flow in the aquifer. This is essentially the reverse of what happens when a well is pumped in a confined aquifer. Until 1976, documented cases of disposal-system failure were rare, but they may increase if underground injection becomes more common and occurs over longer periods of time. Underground injection in the vicinity of old unplugged wells, the locations of which are often not known, may lead to the upward leakage of the wastes.

These sources estimate that there may be more than one million unplugged, unlocated wells in North America that were used for purposes other than disposal. Another hazard associated with the practice of waste injection but only indirectly linked to groundwater contamination is the triggering of earthquakes due to increased pore-water pressures along fault lines. The most publicised occurrence, at the Rocky Mountain Arsenal in Colorado, occurred after injection of chemical-warfare waste, using a 12,000 foot well, into fractured pre-cambrian rocks. The earthquakes thus caused stopped when injection ceased. Other problems may arise if an injection well becomes clogged with injected suspended solids, corrosion products, or precipitation forming between injection fluid and connate water in the aquifer. Insufficient treatment of injected water may lead to bacterial clogging, and swelling of the mineral constituents of the geologic formation. The Report to Congress stated that injection wells can cause groundwater contamination through the following mechanisms:

- direct injection into drinking-water aquifers
- leakage into potable aquifers due to well-construction faults or failure
- leakage through confining beds by hydraulic fracturing or insufficient thickness

- displacement of saline water into potable aquifers
- migration to potable water zone of the same aquifer.

The report concluded, however, that properly constructed, sited, and monitored waste injection wells can be operated with little danger of groundwater contamination.

Disposal of Wastes from Animal Feedlots

Groundwater contamination by the manure from high-density animal-feeding operations is a relatively new phenomenon. In addition to the threat to groundwater from the vast quantities of manure thus generated, there is also a threat from the food additives, such as hormones and antibiotics, which the manure may contain. As recently as 1977 it was claimed that feedlots pose no danger of groundwater contamination in the southeastern states due to the soil's ability to denitrify the wastes.

Feedlots for poultry and hog farms in Arizona, California, Nevada, and Utah were thought to lead to water-quality problems from bacteria, viruses, nitrate concentrations and with general colour, taste, and odour. The main beef-raising areas, however, are the Corn Belt and the High Plains. Until the 1950s and 1960s, beef was raised on pasture, and the wastes were easily assimilated naturally and posed no problem.

The need for more meat at reasonable cost led to the establishment of feedlots, with capacities of between 1,000 and 50,000 head of cattle. During the four to five months each animal spends in the lot, it produces 0.5 tons of manure (dry weight). This can lead to high nitrate concentrations in the groundwater under some conditions. Where the feedlots are located in areas with a deep water table, the risk of contamination is minimised. The main poultry-rearing region is the Mid-Atlantic and South, in Delaware, Maryland, Virginia, Arkansas, Georgia, and Mississippi, for hogs it is the Midwest, and for sheep the Far West.

Of the potential contaminants in manure, nitrate is the most important, as it is soluble in water and its concentration remains unchanged while passing through the soil. The other contaminants from manure bacteria and phosphate are highly attenuated by the soil and pose less threat. Feedlot management, including stocking rate, density, and manure removal, plays an important role in the protection of groundwater quality. Heavy manure accumulations may produce an impermeable mat, which in turn produces anaerobic conditions that favour denitrification. Thus nitrate is volatilisied, and little infiltration takes place, minimising the nitrate-contamination potential. Concentrated animal feeding operations are regulated by the Federal Water Pollution Control Act Amendments of 1972 and may require a permit as a point source under the National Pollutant Discharge Elimination System (NPDES).

Contamination from Radioactive Sources

Radionuclides from natural sources, radioactive fallout, nuclear testing and accidents, nuclear fuel cycling, commercial and industrial products and wastes pose a threat to groundwater quality. Almost all groundwater contains some amount of naturally occuring radioactive substances. Types and levels vary geographically and depend on local geology. The highest concentrations of natural radiation of groundwater are often found near granite and phosphorus formations.

Radioactive wastes from the production and generation of nuclear fuel and radioactive materials can be categorised into five types: spent fuel, high level wastes, transuranic wastes, low level wastes and uranium mill tailings. Spent nuclear fuel is irradiated fuel resulting from nuclear power plant operations and is highly radioactive. Most of these wastes are presently being stored in pools at reactor sites until underground repositories are found for permanent disposal.

High level wastes result from the initial processing of spent fuel and must be stored carefully in specially constructed facilities. Presently these wastes are being temporarily stored at four utility and defence sites awaiting permanent disposal.

Table: *Current and Projected Quantifies of Radioactive Waste and Spent Fuel (September 1983)*

	Volume of Waste (1000 m)		
Type of Waste	***1983***	***2000***	***2010***
High Level Waste			
Commercial	2.3	.436	3.17
Defence	304	294	269
Transuranic Waste			
Commercial	—	4.6	49.4
Defence	246	340	397
Spent Fuel			
Commercial	4.626	19.378	33.258

Contd...

	Volume of Waste (1000 m)		
Type of Waste	***1983***	***2000***	***2010***
Defence	0	0	0
Lower Level Waste			
Commercial	1,020	3,393	5,415
Defence	2,060	3,720	4,710
Inactive Uranium Mill			
Tailings	14,314	23,700	23,700
Active Mill Tailings	96,500	188,800	280,300

(DOE/Defence tailings are about 35% of these totals.)

Source: Adapted from *Spent Fuel and Radioactive Waste Inventories, Projections and Characteristics*. U.S. Department of Energy, 1984.

1. No reprocessing.
2. Found primarily at inactive uranium mills located in western United States. DOE responsible for stabilisation and control of mill tailings in safe and environmentally sound manner under the Uranium Mill Tailings Radiation Control Act of 1978. Anticipated cleanup completion in late 1980s.
3. Located at the 16 active licensed uranium mills operating in 1983, all in the western United States.

Deaf Smith County, Texas, Yucca Mountain, Nevada, or Hanford, Washington. The site is scheduled to be complete by 1998 (Environmental Reporter, 1986).

Groundwater contamination by high level radionuclides has been reported in a few instances, largely due to the improper storage of wastes. For example, in the early 1950s, one hundred gallons of high level radioactive wastes leaked from a tank at the Savannah River defence storage site, contaminating some groundwater. In 1973 a 115,000-gallon leak of high level radioactive wastes occurred at the Hanford Nuclear Reservation.

No groundwater contamination resulted, although the soil beneath the site remains radioactive. Another extensive leak of high level radioactive wastes occurred in 1956, spilling 450,000 gallons of high level radioactive wastes in the ground at Hanford, Washington, but with no serious contamination resulting.

Transuranic wastes contain alpha-emitting radionuclides of atomic numbers greater than 92. They result primarily from fuel reprocessing and the manufacture of plutonium containing products (OTA, 1984). These wastes take a long time to decay and require long-term isolation for proper disposal. Presently, there are a total of 7 sites where transuranic wastes may be deposited (OTA, 1984).

Low level wastes exist in all phases and consist of a wide range of waste material. Most of the radionuclides in low level wastes are short lived and have low radioactivity, but are produced in potentially hazardous amounts.

These wastes are generated by hospitals, laboratories, industrial plants, nuclear power plants, and government and defence laboratories and reactors. In the past, low level wastes were either buried in trenches at government sites or dumped out at sea. Now all commercially produced low level wastes are permanently disposed of in one of the three commercial shallow nuclear burial facilities located in Beatty, Nevada, Barnwell, South Carolina, and Hanford, Washington, the major site being in Hanford.

The Department of Energy and the Department of Defence also maintain a number of sites operated by private contractors for low level disposal (OTA, 1984). Shallow burial of low level wastes present a great potential for contamination. Groundwater may erode the structure of the burial site thus providing avenues for contamination to seep. A recent study found that water has come into contact with wastes in burial trenches at 6 of the 11 major low level governmental and commercial burial facilities.

Uranium mill tailings are earthen residues, often in the form of sand, left after uranium is extracted from ores. Disposal commonly takes place in shallow burial grounds located near refineries. Remedial actions at inactive mill tailing sites are to be conducted by the Department of Energy under the Uranium Mill Tailings Radiation Control Act (OTA, 1984). According to the League of Women Voters (1985) twenty-four abandoned disposal sites are scheduled for clean up by DOE in Arizona, Colorado, New Mexico, North Dakota, Oregon, Pennsylvania, South Dakota, Texas, Utah, Washington, and Wyoming. Of the 28 active uranium mills licensed in the US, 21 reported some degree of groundwater contamination. Remedial action has been taken at 16 of the sites so far.

Other Sources of Contamination Accidental Leaks and Spills

Accidental leaks and spills can occur because of breaks in pipelines, leaks in underground storage tanks, spills from tanker trucks, and other accidents involving the transport and storage of materials. They commonly occur at airports, industrial sites, along highways, railroad sites, gas stations, and refineries.

Small hydrocarbon spills may be absorbed by the unsaturated soil zone, but large ones can reach the water table aquifer and float on the surface. Very small concentrations of petroleum products will render groundwater unfit to drink because of its objectionable odor and taste and also pose a threat to health. Other chemicals have toxic properties. Ammonia, for example, increases the nitrification of groundwater, and acids speed up the solution of heavy metals and soil solids.

The U.S. General Accounting Office (1984) reported that leaks from underground storage tanks (USTs) have been reported in all 50 states; many states claim that this source is their leading cause of underground contamination. There are an estimated 2.5 million underground storage tanks across the U.S., containing more than 25 billion gallons of nonhazardous liquid products, and an additional 2,031 hazardous storage and treatment tanks containing an estimated 13.8 billion gallons of substances (OTA, 1984).

Data was collected at the state level on the numbers of release incidents for EPA's (1986) report *Summary of State Reports on Releases for Underground Storage Tanks*. State files identified 12,444 release incidents that had occurred across the country as of 1984. Even though this data represents the most comprehensive currently available information on leaking USTs, these numbers cannot be regarded as representing all storage tank releases. Differences in state enforcement and reporting procedures have resulted in inherent biases in the data. The results of the study showed that the greatest number of incidents occurred in the Northeast, Mid-Atlantic, and Great Lake states. This is probably due to the more corrosive conditions of the soils, the higher tank population, and perhaps more efficient procedures for reporting and documenting leaks. Soils became contaminated in 68% of the cases, and groundwater contamination resulted from 45% of the releases. It was found that approximately 65% of the incidents originated from retail gasoline stations, and only 3% of the releases involved chemicals other than petroleum fuels. Ninty-five percent of the leaks were reported to have taken place from operating facilities. The mean tank-leakage age was found to be 17 years for all of the tanks. Eighty-one percent of the tanks were steel and the remaining 19% fibreglass. The most commonly reported cause for release was corrosion and structural failure. The NAS estimated that 16,000 spills occur annually during transport operations (OTA, 1984). In 1981 alone, 10,072 spills from both oil and hazardous chemicals, totalling 19.6 million gallons were reported (OTA, 1984). The Department of Transportation reported that more than 9,000 spills of hazardous materials occurred during transportation in 1981 and 6,500 in 1982 (OTA, 1984). Emergency measures to deal with spills often consist of flushing the spilled compound away, thus permitting or even assisting in the contamination of groundwater. Flushing usually reduces the immediate threat of fire and explosion.

Such practices are more likely to create surface water pollution than groundwater pollution, as most storm sewers and ditches flow into surface water. If the spill is large enough and is not properly cleaned up, however, groundwater contamination can result. Under the Clean Water Act, Spill Prevention Countermeasure Control (SPCC) plans have been required of every facility since 1975 for spills to the ground. The National Contingency Plan, under CERCLA, effective December 1982, details cleanup procedures for spills of oil and hazardous materials on land, into surface water, or into groundwater. In addition, releases of quantities of oil or hazardous material in excess of stated amounts must be made known to the National Response Centre. Many of the industrial trade associations have researched cleanup and containment and have published manuals. For example, the American Petroleum Institute manual for the cleanup of underground spills (API, 1980).

Pipelines are used to transport a variety of fluids, but the most common one is petroleum. In 1972 there were 174,000 miles of petroleum pipelines that transported a total of 8.5 billion barrels of petroleum products. In 1976 the same amount of pipeline transported 9.63 billion barrels of petroleum products. In 1971 there were 308 interstate pipeline accidents, involving a loss of about 245,000 barrels of liquids. Pipeline failure due to external corrosion caused 33% of these accidents, and damage to pipeline from excavating machines caused an additional 22%. In 1980 there were 275 accidents; in 1981, 198 accidents. The decrease in incidents may be due to extensive regulations from the Department of Transportation regulating transportation of liquids by pipeline. Of the 4,112 accidents that occurred between 1968 and 1981, 1,372 were due to corrosion and 1,101 to pipeline 83 ruptures. No information was given about the number of barrels involved in the incidents in 1980 and 1981. In some regions deliberate leakage and spillage of contaminants, so-called midnight dumping, has occurred. This source of contamination, which is a criminal offence, is considered to be a major problem in New England.

Agricultural Activities

Irrigation return-flow; the use of pesticides, fertilizers, and manure; and changes in vegetative cover have all been known to cause changes in groundwater quality. Contamination of groundwater by nitrates is one of the largest sources of nonpoint pollution in the United States. The USGS has collected nitrate data over the past 25 years from approximately 124,000 wells nationwide. Its data revealed areas of groundwater that exceed EPA limits for nitrate in almost every state. The study suggests that a large contributing factor to elevated levels of nitrates is agricultural activities such as the use of fertilizers, irrigational practices, dryland farming, and livestock wastes.

Other contributing factors mentioned are septic systems, land disposal of wastes, industrial wastes, and various natural sources. An EPA study of rural water conditions (1986), however, found that only 2.7% of the nations rural wells had nitrate levels exceeding EPA's limits. In the North Central region, 5.8% and in the Northeast region, 0.3% were reported in violation of standards. Hallberg (1985) determined that since the 1960s, an increase in the use of nitrogen fertilizers has been paralleled by a similar increase of nitrates in groundwater. Fiftyfour million tons of commercial fertilizer were used between 1980 and 1981, 48.7 million tons between 1981 and 1982, and 42.3 million tons between 1982 and 1983 (OTA, 1984).

***Table:** Summary of Nitrate-Nitrogen Concentrations in Groundwater, by State*

State	*No. of Wells Sampled*	*% of Wells with NO_3-N 10 mg/l*
Alabama	244	0.0
Alaska	1,305	2.4
Arizona	4,164	13.9
Arkansas	2,436	3.9
California	2,732	10.1
Colorado	5,492	5.7
Connecticut	348	2.3
Delaware	165	9.1

Contd...

State	**No. of Wells Sampled**	**% of Wells with NO_3-N 10 mg/l**
Florida	3,140	2.0
Georgia	1,137	0.5
Hawaii	164	0.0
Idaho	1,806	1.7
Illinois	359	8.4
Indiana	650	1.4
Iowa	4,088	5.0
Kansas	1,140	20.0
Kentucky	3,227	4.2
Louisiana	3,177	0.6
Maine	147	2.0
Maryland	1,215	6.8
Massachusetts	414	1.2
Michigan	1,108	1.1
Minnesota	1,655	9.3
Mississippi	1,701	0.2
Missouri	2,165	2.1
Montana	2,821	3.8
Nebraska	2,326	9.3
Nevada	465	0.9
New Hampshire	69	1.4
New Jersey	1,385	1.4
New Mexico	4,685	2.9
New York	2,491	11.0
North Carolina	908	0.8
North Dakota	7,387	4.6
Ohio	339	2.6
Oklahoma	1,724	11.8
Oregon	685	1.2
Pennsylvania	4,326	5.9
Puerto Rico	79	2.5
Rhode Island	171	36.3

Contd...

State	***No. of Wells Sampled***	***% of Wells with NO_3-N 10 mg/l***
South Carolina	557	0.7
South Dakota	1,996	6.7
Tennessee	109	0.9
Texas	36,196	9.4
Utah	3,301	2.0
Vermont	73	1.4
Virginia	762	0.8
Washington	1,158	4.3
West Virginia	954	0.5
Wisconsin	2,727	3.6
Wyoming	1,477	3.8
Total	123,656	6.4

Source: Madison and Brunett, 1985.

Nitrogen accounts for 6.1% to 20.4% of these weights. Much of the applied nitrate is not utilised by crops and is subsequently lost through leaching.

Less nitrate-nitrogen is available for leaching in the loamy finesand soils of subirrigation systems than in furrow and sprinkler systems. Thus, subirrigation systems provide a method for reducing solute concentrations in irrigation return-flow. Irrigation only contributes further to nitrate leaching. Less than 1% of the total cropland in the Northeast is irrigated; as a result, nitrate contamination is not that much of a problem. In the West, however, due to the highly saline soils, irrigation often exceeds soil requirements. Dryland farming in the northern Great Plains does not utilise irrigation, and crop plants grown in the area often have lower evapotranspiration rates; therefore water may move down through the soil picking up salts and thus contaminating the groundwater. In some cases this salty water produces saline seeps.

In 1983, 12 states experienced groundwater contamination from pesticides applied to agricultural land. In 1984, a total of 12 pesticides were discovered in the ground waters of 18 states. By 1985, 23 states discovered groundwater contamination from 17 different pesticides. Although agricultural operations alone account for 69%-72% of all pesticide use, contamination by pesticides can also occur through spills, leaks, lawn and garden application, and disposal practices (OTA, 1984).

Table: *Pesticides Found in Groundwater as a Result of Normal Land Applicationå*

Pesticide	***Use****	***State(s)***	***Typical Positive, ppb***
† Total of 17 different pesticides in a total of 23 different states.			
E* H = herbicide; 1 = insecticide; N = nematicide			
Alachlor	H	MD, IA, NE, PA	0.1-10
Aldicarb (sulfoxide& sulfone)	I, N	AR, AZ, CA, FL, MA, ME, NC, NJ, NY, OR, RI,TX, VA, WA, WI	1-50
Atrazine	H	PA, IA, NE, WI, MD	0.3-3
Bromacil	H	FL	300
Carbofuran	I, N	NY, WI, MD	1-50
Cyanazine	H	IA, PA	0.1-1.0
DBCP	N	AZ, CA, HI, MD, SC	0.02-20
DCPA (and acidproducts)	H	NY	50-700
1,2-Dichloropropane	N	CA, MD, NY, WA	1-50
Dinoseb	H	NY	1-5
Dyfonate	I	IA	0.1
EDS	N	CA, FL, GA, SC, WA, AZ, MA, CT	0.05-20
Metolachlor	H	IA, PA	0.1-0.4
Metribuzin	H	IA	1.0-4.3
Oxamyl	I, N	NY, RI	5-65
Simazine	H	CA, PA, MD	0.2-3.0
1,2,3-Trichloropropane	N (impurity)	CA, HI	0.1-5.0

Source: U.S. EPA, 1986b.

EPA believes that with increased groundwater monitoring more cases of contamination involving additional agricultural chemicals will be discovered (EPA, 1986b). The pesticide dibromochloropropane (DBCP), now banned in the United States, adversely affects human health and causes reproductive dysfunction in laboratory animals when given at high concentrations. There is no direct evidence for adverse effects on human health at the concentrations found in contaminated groundwater (CEQ, 1981a). Although not really an agricultural activity, suburban lawn care can cause groundwater contamination (Miller et al., 1974).

Mining

The formation of acid mine-drainage water has been described above. In areas where coal mining has been carried out over long periods of time, in Appalachia and Pennsylvania, for example, it is estimated that due to its low pH, the groundwater would be unusable for decades after the cessation of mining activities. Metal-mine leachate has increased manganese in wells in Washington, caused arsenic poisoning in cattle in Idaho, and increased the radioactivity of groundwater in Wyoming.

Highway Deicing Salts

Many of the northern states use large quantities of salt in combination with abrasives such as sand on icy and snowy highways. The salt-solution runoff can percolate through adjacent soils or otherwise find its way into the groundwater via storm drains. Due to its low cost, salt is often stored in uncovered piles. Precipitation dissolves the stored salt, which may then infiltrate shallow aquifers. Large amounts of salt are used on northern highways each winter. Over 12 million tons were used in the winter of 1978-79. During 1982 and 1983 at least 9.35 million tons of dry salts and abrasives and an additional 1.78 million gallons of liquid salts were applied to highways (OTA, 1984). During a typical winter approximately 17.6 tons of salts are used per lane per mile (OTA, 1984). The amount of salt used depends, of course, upon the severity of the winter, upon the amount of abrasives used, and upon driver education. Nevertheless, salt usage in the Northeast has resulted in chloride contamination of many drinking-water wells adjacent to highways.

The dangers to public health of high salt concentrations in drinking water are not well known. Water is considered contaminated with chloride when the concentration of chloride exceeds 250 mg/litre; concentrations above that level leave a salty taste in the water. High concentrations of sodium have been linked with hypertension and have been implicated in the exacerbation of certain liver and kidney problems. The American Heart Association has suggested a maximum level of 20 mg/liter in drinking water. High concentrations of salt can also be severely detrimental to plant life.

Atmospheric Contaminants and Acid Rain

Very little is known about the effects of atmospheric contaminants and acid rain on the quality of groundwater. In Michigan chromium in the dust from an industrial plant was leached into an aquifer and contaminated a well. Hubert and Canter (1980b) have reviewed the literature concerning the relation between acid rain and groundwater, and they conclude that acid rain may react with soils and surface waters in such a way as to increase the leaching of metals and nutrients. Thus, the potential for groundwater contamination is present and should be investigated further, especially the potential for lead and cadmium contamination via leaching because of their long environmental residence time.

Infiltration of Surface Water

Aquifers are often in hydraulic connection with bodies of surface water. Under certain conditions, polluted surface water from lakes or streams can percolate into a water-table aquifer, thereby degrading the quality of the groundwater. Also, groundwater development near a body of surface water may draw contaminated water into the aquifer.

Development of Groundwater

Pumping of water wells may bring water of lesser quality into their zone of influence. In inland areas serious problems may be caused by the migration of water from saline aquifers through leaky aquitards into potable supplies. In coastal regions extensive groundwater development has led to saltwater intrusion. Saltwater encroachment is an important source of groundwater contamination in many populated coastal regions, particularly in parts of California, Texas, Louisiana, Delaware, Florida, and New York. Saltwater occurs naturally in the aquifers in coastal areas, and pumping fresh water from wells may cause saltwater intrusion into potable aquifers, rendering them unfit for use. In 1981 the USGS estimated that 21 billion gallons of water per day, or 26% of all groundwater withdrawn, is in excess of natural recharging capabilities. In 1985 the EPA reported 19 states having problems with saltwater intrusion from deep saline aquifers or from the ocean. Attempts to prevent such contamination include the injection of freshwater "barriers" to reverse the hydraulic gradient in the aquifer so that the flow is toward the sea rather than toward the pumping wells. This method has met with some success in the Los Angeles area. Land subsidence has occurred in some areas of the United States because of excess withdrawal of ground waters.

A dramatic example of this occurred over a period of 37 years in San Joaquin Valley, California, where an area of land subsided nearly 29 feet.

Improper Construction and Maintenance of Wells

Faulty, corroded, or ruptured well casings, the linking of two aquifers by a well screen, or open wells can cause contaminants to move from one aquifer to another. Sources of contamination from the three main categories described above have been documented from all parts of the country. Whether they cause localised problems or whether the groundwater-contamination problems fall into larger regional patterns has been a matter of dispute. Our conclusions, based on a study of the documented cases of groundwater contamination available, together with examples of known occurrences of groundwater contamination for selected states. It may be said, however, that groundwater contamination is not usually direct, and in most cases the contaminant must pass through the soil layer. Attenuation of contaminants has been thought to be greater in the unsaturated zone because of the greater possibility of aerobic degradation, adsorption, complexing, and ion exchange of organics and inorganics. Recent investigations on anaerobic metabolism, however, have discovered that many polluting compounds can be significantly degraded under reduced conditions. Dilution, buffering, precipitation, reduction or oxidation, mechanical filtration, volatilisation, sorption, and other processes may operate within the aquifer to purify chemical wastes.

3

Groundwater Site Characterisation

In the field of management, the "Principle of Uniqueness" describes the notion that certain problems require unique solutions. It is an understanding that has resulted from unhappy experiences in attempting to apply universal solutions to the complex problems that exist in the real world. Rather than applying a structured set of prescriptive dictums or maxims to solve a recognised problem, this approach focuses on a process for achieving satisfactory outcomes. Recognition of the importance of multiple perspectives and non-linearity in real world problems has been the catalyst for the systems movement, and it applies directly to the problem of characterising groundwater flow and contaminant transport phenomena. A basic tenet of the systems movement is that there are multiple dimensions of significance (or attribution) which must be considered when evaluating a problem.

The first dimension for consideration is the level of emergence; which, according to Checkland, roughly translates into: "The system is more than the sum of its parts." The system is in some way purposeful. A system, whether human or natural, must be defined and considered as a whole entity which cannot be reduced to unconnected component activities and structures. The second dimension is hierarchy, which builds on the concept of emergence. The underlying concept is that systems are built up of smaller entities which are themselves wholes. Emergent properties (which relate to purposeful conditions such as niches or functions) denote the hierarchical level. Integral to systems thinking is the belief that a problem must be analysed at the same level as the problem itself, and not below it. This means that a problem involving a complex human or natural system must be analysed by an inclusive examination of multiple perspectives as defined by Linstone.

When considering groundwater issues, the level or degree of site characterisation must be sufficient to identify not only the physical and chemical conditions present, but also be address the legal and sociopolitical aspects inherent in remediation (such as site history, water rights, etc.). Economics is also an important consideration; characterisation is linked directly to cost. Whether a cost is justified is based upon the regulatory climate, as well as the magnitude of the actual and perceived risk present. Perceived risk and regulatory climate are interrelated and best characterised by multiple perspectives that must be considered before site characterisation can be successfully completed.

Characterisation of Groundwater Systems

Groundwater systems are nonsteady state, dynamic and frequently subject to spatial variation. The basic physical framework is the site specific geology. Groundwater flow is dependent on both spatial and temporal factors including variations in hydraulic conductivity, gradient, porosity and thickness of stratigraphic units. Temporal variability occurs as the result of seasonal recharge, variable withdrawal of groundwater, and intermittent contaminant releases. Groundwater flow is both cyclical and noncyclical in time.

Contamination, if present, adds additional temporal and spatial uncertainty to the description of the groundwater system. As sources of contamination vary in time, the nature of the rate and extent of contamination in the groundwater changes. Chemical reactions and biotransformations also influence the kinetics of contaminant migration. For example, reductive dehalogenation is a well-documented biotic pathway for the breakdown of certain chlorinated organic substances. Abiotic pathways for chemical reaction and transformations have also been identified. Obviously, kinetic properties must be defined for the groundwater contamination problem to be properly characterised. Site characterisation is both spatial and temporal, and it requires identification of steady state and nonsteady state influences. The complexity of site characterisation and its relationship to contamination and public health concerns can be illustrated according to the procedure described by Checkland.

The hierarchy illustrates the range of issues and perspectives that must be spanned if a groundwater remediation program is to be successful in meeting the mandates set by our regulatory and legal communities. To accomplish this, characterisation of sites undergoing remediation must include an explicit approach to identify and describe the unique geological and other natural processes present, but also, to identify and accommodate what Linstone has described as the "multiple perspectives of the relevant actors" involved. He suggests that such perspectives fit into three categories: technical, organisational, and personal. These terms can be defined in the context of Checkland's systems hierarchy.

While each perspective is essential to the success of a groundwater remediation project, it is risky to assume site characterisation is complete until the technical perspective is satisfied. The problem of groundwater site remediation must be accurately defined and described at each relevant level of emergence, which at Levels 1 through 6 are principally technical in nature. Preoccupation or preference for an organisational perspective (i.e., assigning it a higher priority) can constrain and inhibit the technical perspective by restricting site characterisation resources (funds, personnel power, equipment) without regard to the technical conditions present and their implication. This short-term focus on fiscal control would therefore be applied at the expense of the risk of long-term financial exposure from liabilities associated with reduced understanding of the site. The organisational perspective, including the need for financial control, is enhanced by having a solid technical foundation. Civil engineering and the organisational response A new field of civil engineering practice has developed during the last 10 years groundwater remediation. Typical civil engineering projects involving environmental controls proceed in steps by an incremental approach.

The first engineering step is the feasibility study, which is an alternative assessment that spans the eight levels of systems emergence described earlier, including the technical and organisational perspectives of authors and intended audience. Once a project is approved for further design development, engineering proceeds in the following steps:

- conceptual design,
- preliminary design, and
- detailed design.

Engineering design occurs at the third level of systems emergence. Design is staged in a convergent, linear, programmable, and predictable sequence. The management decision process of incrementalism fits conveniently into the framework of engineering design. Applied inappropriately, this can inhibit the divergent aspects of site characterisation by restricting funding, resources, and time allocated to this important activity while prematurely emphasising the process of design.

Martin has described the "Convergent-Divergent Cycles" associated with managing technological innovation. Convergent thinking involves the inherent need to drive on to the solution of a problem, and it can involve a number of cognitive styles and tools, including morphological analysis, relevance trees, and value analysis, among others. Divergent thinking involves the generation of new ideas based on brainstorming, analogies, lateral thinking, random juxtaposition, and others. The process of bringing convergent and divergent thinking together is the fundamental task of the research and development manager such as a project manager. The process of integration involves explicit steps: problem definition, divergent thinking, convergent thinking, evaluation, and a criteria for iterating back to divergent thinking. This is a learning system.

For many years, geologists have used this approach to define and describe geological conditions at a given location. Formal utilisation of this approach brings elements of divergent thinking to bear on issues of complexity in natural processes, which is inherently part of geological field investigations. As field work progresses and information gathered, divergent thinking is used to generate as many interpretations as consistent with the data. Tests are then formulated to check the various interpretations by data already obtained or by checking against future data gathered with this in mind. Many possible scenarios can be considered, most will be abandoned and the final interpretations often bear little similarity to those postulated early in the process. The process of multiple working hypothesis may seem inefficient; it is not. It is a highly interactive learning process that has been widely utilised in other applications for many years.

It is a very cost effective technique when applied properly. It is not, however, widely taught or used by the civil engineering community. This situation is easily resolved by structuring the project team to include experienced and appropriately qualified geologists, chemists, statisticians, and managers. An important goal of the site characterisation project team is to provide the engineering team with sufficient information concerning contaminant rate and extent to design a cost effective approach to remediation. Engineering, as a formal process, begins at the feasibility study once a project has been identified. A preliminary or feasibility study is performed to consider, in detail, the implementation of alternatives and associated costs. The utility of a feasibility study related to groundwater issues is directly related to the detail and extent of site characterisation. Given a poor characterisation, no engineer can perform an adequate feasibility study, nor can any organisational perspectives be appropriately addressed. The integration of these tasks has to be one of the main objectives assigned to the project manager. Penite remediation initial engineering approach

From 1940 until 1971 sodium arsenite (Na_3~AsO_4~$12H_2O_2$~) was manufactured under the trade name of "Penite" at the former Pennwalt plant in Tacoma, Washington. Penite was used as a herbicide. It is very soluble in water and contaminated groundwater beneath four acres of the facility. In 1986, the state of Washington entered negotiations with the Pennwalt Corporation to investigate and, if appropriate, to remediate the contaminated groundwater. Consulting engineers interviewed during the selection process articulated what could be considered an organisational perspective this effort would not become "a research project."

Divergent thinking in this case was strictly bounded for the sake of economy. When this occurs, the potential for project success diminishes significantly. Environmental investigations that began in 1987 led to the conclusion that remediation was indeed required and several options for cleanup were generated. However, these options were identified using a process managers call "bounded rationality." The concept of bounded rationality refers to the tendency of managers to select less than the best objective or alternative by engaging in a limited search for viable alternative solutions, using inadequate information, and controlling the factors influencing their decisions. Normally, the rationale for this short cutting approach to save money is supported by an underlying fear that the more we know, the more it will cost and the longer it will take. Because the scope of alternative assessment was constrained in this manner, the only apparent remediation strategies for the Penite site all involved containing the entire volume of contaminated water beneath the site forever.

Early in the investigation the engineers concluded that groundwater beneath the site simply was untreatable. This conclusion was reached after expending very little time and expertise investigating how dissolved arsenic might be treated for removal from contaminated groundwater. And, because very limited data on physical aquifer properties were collected and analysed, the consultants failed to identify a severe fatal flaw in their initial containment strategies: that an impermeable layer believed to be uniformly present beneath this site simply was not. Further, limitations in the data base forced unrealistic assumptions to be made and a high degree of uncertainty in the prediction of contaminant hydraulics. The proposed containment options ruled out any future use of four acres of prime industrial waterfront. What's more, there was no evidence or certainty that the performance objective, to reduce the net flux of arsenic in groundwater leaving the site by at least 80%, could ever be met. Corporate senior management was not satisfied with this risk.

The Penite Remediation Using a Systems Approach

In late 1988 and early 1989, Pennwalt (now Elf Atochem) reorganised the remediation project incorporating a technology research and development project management approach. The intent was to fully integrate what has been characterised as "the divergent thinking scientists (brainstorming, wild ideas, lateral thinking, etc.) and the convergent thinking typical of engineers and some managers (morphological analysis, relevance trees, value analysis, etc.)." Such assimilation of different cognitive styles was necessary to avoid the pitfalls and constraints imposed by convergent thinking prior to full consideration of possible situations and scenarios. Furthermore, since research involves a structured approach to learning, this project was structured as a learning system with information flows forward to formalize research. A systems approach was deemed more appropriate to the groundwater problem at hand because it could link together pertinent disciplines in a logical manner to create hardware, operational procedures, and subsystem linkages to accomplish a predetermined goal; and from the organisational perspective, would be very cost effective in the long term.

According to Martin, creativity is "an essential component of complex sociotechnical problem-solving." Therefore, it must be included in the management of groundwater remedial action projects. Technological perspectives include both systemic and intuitive approaches. Systematic thinking involves empiricism and rational, analytical formulations. The successful project management approach must be able to accommodate personal perspectives, including creativity, autonomy and challenge, as well as tolerate disagreement and provide reward for successful accomplishment. This organisational perspective stimulates and sustains a learning system that integrates the diverse project team. A new project team was formed that included internal technical and managerial expertise supplemented by several outside scientists. A scope of work was formulated and additional staffing needs identified. Rather than select a single, multi-role consultant, experts were selected from various firms and integrated into a project team.

Because time constraints imposed by the regulatory community were very restrictive, several parallel tasks were essential. While additional site characterisation information was developed, a three-dimensional groundwater contamination transport model was developed. A comprehensive evaluation of hydraulic containment and control, as well as chemical engineering processes applicable to arsenic separation/removal, also was initiated. Analysis conducted during these new site investigations indicated that the containment option proposed by the initial engineering approach could not have met the site specific clean-up goals. The site had not been completely characterised and its uniqueness was not fully recognised. To understand why the containment option was inappropriate, a little site background information is required.

In the mid to late 1920s, an artificial waterway was dredged along the bed of Hylebos Creek location of the Penite site to allow marine shipping access to Tacoma's Commencement Bay in Washington State. Dredged materials consisted primarily of fine-grained silts and sands, and approached thicknesses of 8 to 12 feet beneath the existing site. Groundwater, recharged by rainfall, occurs in this deposit. The saturated thickness of the deposit is in the range of 8 to 10 feet. The former land surface was largely composed of silt and clay-sized particles with organics, such as vegetation and peat intermittently present in its thickness. This stratum is less permeable to groundwater and was described as an aquitard by the initial investigators.

This aquitard was an important component of passive containment for the arsenic contaminated groundwater. Since the aquitard was present in all 28 initial borings at the site, it was assumed to be a uniform feature. Unfortunately, the uniformity of geological features is very scale dependent. In fact, the geological processes that increased the permeability of the layer should have been postulated by geologists familiar with sedimentary facies present at this site based on a review of U.S. Geological Survey maps, U.S. Army Corps of Engineers charts and photographs from the early 1920s. Additional borings confirmed the presence of vertical migration pathways through the aquitard. Since a pathway existed between the contaminated uppermost aquifer and a deeper uncontaminated aquifer, the remedial objective of reducing the net loading of arsenic to the adjacent waterway by 80% could never have been assured by passive containment.

Kriging, a geostatistical technique employed by the mining industry for many years, was used to quantitatively evaluate site characteristics. Kriging is a statistical procedure which can be used to estimate spatial values such as hydraulic conductivity and water table elevations. It is a complex weighted averaging technique that takes into consideration the spatial relationships between data points and a grid network. Spatial data values are extrapolated at each gridnode with minimum error variance. A map of variances can be generated which reveals regions of high or low data confidence. Areas of low confidence require additional data: when the map of the site indicates high confidence, little or no additional data acquisition is warranted.

Kriging can be used as an objective, hypothesis-testing mechanism during site characterisation. It is also a useful tool for contouring spatial data. Based upon these results, site specific parameters were incorporated into the mathematical description of groundwater contamination beneath each site. This model was used to establish more refined strategies for achieving the remediation goals. It was determined that removal of concentrated arsenic in soil, reduction of onsite recharge, and placement of a steel barrier wall would reduce the discharge of arsenic in groundwater by at least 80%, thus meeting clean-up performance goals set by the regulatory community. The problem of finding treatment methods for removing arsenic dissolved in groundwater was stymied when it was discovered that high concentrations of silica were dissolved in the groundwater which resulted in a troublesome gel when conventional approaches to metals removal were attempted. Given this difficulty, the initial engineering team resorted to containment as a solution.

Decision-making by this type of process is known as satisficing. It can be problematic in certain circumstances. Interestingly, when the new project team realised that the containment option was severely flawed and turned its attention to treatment technologies, the integration of convergent and divergent thinking allowed it to look beyond the removal technologies then known by the environmental community. The team looked into creative solutions involving synergies such as separation technologies used in the chemical process industry. As a result, the project team's technology development program led to the implementation of a hybrid process for arsenic removal from groundwater. This hybrid utilises a two-step precipitation that takes advantage of a sulfide reaction and a subsequent iron precipitation step. The sulfide precipitation step has long been used for brine purification by chloralkali process engineers and precipitation by ferric chloride is currently used by environmental engineers for metals separation in Europe; the combination is, therefore, both synergistic and leveraged.

The remediation technology package was a set of integrated subsystems. The package was approved by the regulatory agencies in discrete steps, although the project team maintained an integrated perspective at all times. While this effort appeared to be incremental, no aspect of the project was ever isolated or considered "stand alone." In late 1991, the Washington State Department of Ecology approved the design of the integrated extraction and treatment system. Construction began in early 1992 and the system startup occurred in early October 1992. Options for commercialising this treatment technology are being explored. At present, the release of arsenic from the site has been greatly reduced; it is hoped that the site will be once again available for industrial use within 10 years.

Lessons Learned

Groundwater remediation projects are highly complex sociotechnical undertakings which must address multiple perspectives as defined by Linstone. Such efforts require the integration of many levels of information and the incorporation of multiple perspectives. Successful groundwater remediation is more than achieving a prescribed clean-up goal; it must be achieved with consideration of cost and long-term effectiveness. Project management approaches must accommodate more complexity than exists in most common engineering projects; site-specific factors make each effort unique. Scientific invention and engineering development must be championed and managed according to recognised needs in a supportive environment.

There are many biases that interfere with this process including a commonly encountered Fear that such projects will become endless exercises in data acquisition and cost. Yet, the tools and techniques exist and are readily available to assure that the right amount of information is obtained and the desired results achieved. Models, properly calibrated with appropriate sensitivity analysis, are important operations research tools. However, simulation models are never complete representations of reality. Models must always be suited to the problem. The cost of information, which includes simulation, must always be economically justified by the value of the decisions involved. The value of restoring contaminated property to industrial use is obvious, and it must be considered as an opportunity cost if not completed. There is value in developing new technologies which might have application elsewhere and, thus, could be marketable.

Since each site characterisation is unique, it i s evident that highly sophisticated models are not always required or appropriate. Site characterisation is itself a kind of model and simple analogues may suffice. The management concept of uniqueness applies. The real world is complex, and risk assessments and feasibility studies are often derived from limited information.

While decisions should never be made mechanically based upon the output of a numerical model or statistical algorithm like kriging, models can establish invaluable range estimates when used in conjunction with appropriate sensitivity analysis. Kriging, or geostatistics, can provide highly illustrative contours of contaminant level (or other geological properties) as well as invaluable contours of the characterisation error associated with a site. Ultimately, both probabilistic and deterministic models used interactively by a qualified team incorporating multiple working hypotheses, are instrumental in effective site characterisation.

The success of any program of technology development is dependent upon the effectiveness of the management process involved. Therefore, if the technical efforts associated with a remedial action are successful, the management process used must have been successful. It is all too frequent to encounter "minimalist" approaches to remediation that rely on satisfying the regulators to limit and control data acquisition. Managers of such projects seem to have forgotten that it is generally easier to regulate than to remediate.

Experience has taught that to protect corporate assets and limit liabilities, it is far better to set the scope of such activities objectively and to aggressively pursue information interactively so decisions can be made according to explicit levels of uncertainty, cost, and potential future risk. This is consistent with a long-term organisational perspective. The goal is a form of humility in decision making concerning remedial alternatives that Etzioni suggest requires decision makers to "stay loose" through the process, thus avoiding the traps of bounded rationality and incrementalism. This point is worth judicious consideration by environmental practitioners in the private and public sectors involved in groundwater remediation projects.

An Environmental Engineering Design

Educational criteria outlined by the Accreditation Board for Engineering and Technology, and supported by practicing engineers, pose a challenge to the engineering educator. Many of these criteria, such as those pertaining to problem solving and communication skills, can generally be met via the refinement of existing courses and exercises. Those concerned with enhancing problem identification and experimental design skills, functioning on an interdisciplinary team, instilling professional ethics, and promoting the desire for lifelong learning are best met by increasing the real-world content of curriculum. Infusion of effective design problems or projects into the curriculum is one way to increase the real-world content. The question then arises: How do we fit more design projects into the already overcrowded curriculum without sacrificing the depth of coverage by existing theory-based material? One answer lies in interactive simulations which are rapidly evolving as vehicles for honing realistic problem-solving skills.

An engineering design project is one type of problem-based learning (PBL) exercise. The PBL approach asks students to self-learn material in the context of a realistic problem, similar to one they would encounter as professionals in the subject area. The instructor's role in the PBL setting is to make appropriate material available to the students and provide real-time feedback in regard to the directions they choose. The motivation behind the PBL approach is that students achieve a deeper understanding when they solve realistic problems by sorting through material of variable relevance.

Examples of multimedia simulations in PBL are abundant in the medical curriculum where many of these techniques were pioneered. If a PBL exercise is to be effectively driven by such a simulation, then the underlying software must have the capacity to: 1) inspire a high level of learner activity and motivation (through subject matter depth and relevance); 2) challenge learners to make interesting and realistic choices; and 3) allow learners to make meaningful changes in the simulation parameters (i.e., let the learners experiment). The approach is well suited for the complex problems faced by medical students learning to diagnose and treat illnesses based on a combination of anatomical, physiological, and historical data. However, examples of this approach are less common in other curricula, and usually involve interactive simulations focusing on relatively specific problems, or on visualisation of complex subject matter.

In this work, we present a PBL simulator capable of delivering a broadly scoped design project. The simulator is an interactive database that administers a design project for environmental engineering students. The primary goal of this work is to advance the development of PBL simulations in the engineering curriculum.

An additional objective is to infuse current contaminant fate and transport research results into an already crowded undergraduate civil engineering curriculum. An important constraint is to achieve both goals without imposing excessive time demands on the instructor. The chapter begins with an overview of software content, features and operation. The subsequent section describes the instructor-based input required by the simulator and the logistics of the design project. Finally, results from a preliminary test of the software are presented and discussed.

Interactive Site Investigation Software

The program for delivering the interactive PBL environment specific to this work is called Interactive Site Investigation Software (ISIS). In the present course, students are taught the principles of contaminant hydrogeology (physical chemistry, soil physics and groundwater hydrology) through conventional lectures and problem sets. About halfway into the course, the students' outside assignments become virtual data collection using the interactive database. More specifically, student teams complete a design project in the form of a Remedial Investigation-Feasibility Study (PJTFS) for a virtual hazardous waste site. Virtual work tasks include drilling, collecting samples, analysing samples and performing field tests, and are designed to link theory outlined in the lectures to realworld situations. The project discussed here was devised for a senior-level environmental engineering course. The ISIS software comprises the simulation database and graphical users' interface (GUI). The database is a set of textbased and binary files generated by the instructor or by mathematical simulations controlled by the instructor. The GUI is a Java (TM)-based (JDK(TM) 1.1, Sun Microsystems) program through which the students manipulate and query the database.

Virtual Field-work in ISIS

The design project is based on a fictitious hazardous waste site referred to as Camp Hazmat. A site plan view is used to describe the layout of this site to the students. The teams logon and start in the ISIS office by selecting a work mode. The majority of the virtual work is completed when the teams go to the field in either the soil-boring or field-testing mode. Otherwise, the teams may go to the laboratory to designate samples for various analyses, research theoretical and practical points in the virtual library, or review documents summarising the team's previously completed work These work modes are discussed further in the following sections. It is important to note that many of the choices presented to the students by: ISIS pertain to time and cost issues. When a team enters ISIS, the simulation clock is initiated and the costing algorithm begins to track all materials and services (including the team members' time).

Soil Boring: The teams begin field work by invoking the soil boring option. ISIS then prompts the team to select the drilling equipment and location (x and y coordinates). The team regulates the drilling schedule and sampling frequency in the control panel. There are three possible drilling strategies: continuous drilling with manual starting and stopping, drilling with automatic stopping at team-specified vertical increments, and continuous drilling with automatic stopping at team-specified depths or events (e.g., encountering groundwater). ISIS accelerates simulation time (16:1) during drilling and other intensive field activities in order to sustain an engaging pace. The advantages of one drilling strategy over another are time and cost related.

With each pause in the drilling, the team has the option of designating a core sample for future laboratory analyses. There is modest cost for the time associated with sample collection and handling, and more significant costs specific to the type of analysis. Thus, budgetary constraints tend to discourage over-sampling. ISIS allows a team to save a soil sample indefinitely without repercusssions. It would be more realistic to develop a chemical-specific sample deterioration routine, and this may be considered in subsequent versions.

Once a borehole has been completed, the team must construct a well. ISIS leads the team through the process of selecting the vertical distribution of well-screening, which fixes future groundwater sampling locations. ISIS allows a maximum of three screened intervals for which the total screened length must not exceed 30% of the well. More screens allow for more vertical sampling opportunity, but the cost of well-construction increases with the number of intervals. Thus, the team must learn to differentiate between scenarios where vertical resolution of the contamination is necessary and when it is extravagant.

Figure *: Soil Boring & Pits*

Field Testing: With wells in place, the field-testing option becomes viable. ISIS prompts the team to identify a well for testing and choose the desired test. Activities range from simply extracting a groundwater sample for later analysis to hydrogeologic testing (e.g., pump and tracer tests). Pump (or drawdown) tests are performed to estimate the permeability of the porous media surrounding the wellscreening. Tracer tests are performed to estimate the degree of dispersion expected during the transport of contaminants in that media. For brevity, only the single-well drawdown option is described here.

Once the team chooses to execute a single-well test, ISIS prompts the team to select a pump. If the well has multiple screens, then ISIS prompts the team to select one for testing. Next, the team learns it must check the initial water level in the well and then begin pumping. Each time the team selects the "sample" button, ISIS registers the sampling time and updated water level. Water levels are calculated using analytical expressions for radial flow to a well. Calculated water levels are reported to the teams as raw data (i.e., the distance from the ground surface to the water level in the well). If at any time the estimated water level drops below the bottom of the screened interval, ISIS prompts the team to select a more appropriate pumping rate.

Based on their knowledge of the well hydraulics, and on experience gained during the project, most teams learn to select reasonably optimal pumping rates so as to reduce the number of costly restarts. It is important to note that these results are based on the students' calculations (not ISIS). This approach is consistent with the current project emphasis on students learning to plan experiments and interpret results.

Analytical Laboratory: Laboratory work is subcontracted using the ISIS chain-of-custody form. Laboratory analyses currently available in ISIS include standard methods for determining chemical concentrations in soil and water samples (EPA methods) and physical soil properties (ASTM methods). Many of the standard chemical analyses screen for numerous compounds. If a given compound is not part of the instructor-generated ISIS database, then it is reported as "not detected." To submit samples to the lab, the team selects from the previously collected soil and groundwater samples, then they check the desired tests for each sample. ISIS will accept a maximum of five analyses for each sample in consideration of sample quantities needed for the analyses. Once the sample chain-of-custody form is submitted, ISIS pauses for an instructor specified amount of time (e.g., regular rate or rush), then mines the ISIS database for the appropriate analytical results. ISIS then e-mails the analytical results to every team member after the next logon.

Virtual Library: The virtual library is an optional link to HTML files containing primers on groundwater hydraulics, physical chemistry, contaminant fate and transport theory, descriptions of the drilling and sampling hardware, and standard analytical methods that the teams will be employing on the job.

Archival Documents: ISIS stores all of the work completed by each team in a series of easily accessible borehole logs, expense accounts, and laboratory reports. The teams access archival documents from within the ISIS office. The reports are especially useful near the end of the project for documenting activities and assembling the project report.

Instructor Setup Required by ISIS

Setting up the subsurface contaminant distribution in an ISIS scenario requires the creation of the ISIS database. There are four main components of ISIS database: 1) soil properties, 2) soil contaminant distributions, 3) hydraulic head distribution, and 4) groundwater contaminant distributions. The database is a set of four-dimensional arrays specifying the location and property or dependent variable value described below. The spatial discretization is the same for all properties and currently comprises a grid of 83 columns, 60 rows and 100 layers. The three-dimensional grid encompasses a subsurface volume that is about 4600 feet long by 1900 feet wide by 100 feet deep. For memory and performance considerations, the horizontal grid is irregularly spaced, and finer where the contamination is placed (where more intense probing is expected).

Soil Properties: Setup of the three-dimensional distributions of ISIS soil properties is simplified through the use of a soil palette. This palette currently contains the 15 soil types and physical properties. Thus, the instructor need only create a four-dimensional array detailing location and soil type, and ISIS will employ the soil palette to extract the relevant properties when queried about a particular location. The instructor must take care to insure that the soil type distribution is consistent with the distributed flow and transport parameters assigned to the simulation models discussed below.

Soil Contamination: ISIS tracks contaminants according to the chemical abstracts specification (CAS number). Any chemical may be used as long as its properties are provided by the instructor in the chemical properties palette. Common chemical mixtures (e.g., fuels) can also be included in the project by assigning fictitious CAS numbers and appropriate aggregated properties.

ISIS soil contaminants are distributed at various locations between the ground surface and the uppermost elevation of the groundwater domain, or water table. The soil contamination facet of the database is oversimplified in that ISIS identifies sources according to instructor-specified shape functions (cones, spheres and cylinders). ISIS only gathers contaminant data if students' queries fall within a specified shape's boundaries. Future versions of ISIS may institute a multiphase flow model to simulate the distribution of liquid spills or leaks in the unsaturated soil zone more realistically.

Hydraulic Head: The groundwater conditions are described by the hydraulic head distribution generated by a public domain groundwater flow model MODFLOW. The computational grid for the MODFLOW simulations is identical to the horizontal soil property grid, but only ten cells thick to expedite computations. Thus, the flow parameters of the MODFLOW layers must be assigned values equivalent to the average of the values for the corresponding ten intervals in the soil properties section of the ISIS database. ISIS requires three binary output files from the MODFLOW simulations. The first file verifies the ground surface elevations input by instructor. The second reports the head distributions (i.e., equipotential surfaces) within the ten layers underlying the ISIS site. The third binary output file is necessary to drive the contaminant transport model described in the following section.

Groundwater Contamination: Contaminant distributions for the ISIS database are generated using the MT3DMS model (v. 3.50). This version allows for the simultaneous simulation of multiple solutes. The instructor must become familiar with the input structure of MT3DMS and create a set of parameters that is consistent with the assigned soil properties. MT3DMS produces a binary output file for each of the simulated contaminants. These files are required by ISIS and must be renamed according to the appropriate CAS number in order to be recognised and linked to the chemical properties palette. The distribution of the contaminant trichloroethene (TCE) within one of the MT3DMS computational layers for a portion of the Camp Hazmat site is plotted. The current ISIS database accesses ten such layers for each of the 15 contaminants.

Data Variability: An important attribute of the ISIS database is the variability that is imparted to any datum being extracted by the students. ISIS operates on each datum using a random number generator. This generator employs an instructor-specified statistical variance to synthesize a normally distributed amount of "noise" for each available physical or chemical property. The magnitude of the variance is dependent on soil and chemical type, supporting the fact that some soils and chemicals area easier to characterise than others. This aspect of ISIS was developed to exercise students' data interpretation skills. For example, students might determine that they need to repeat an observation several times before they can differentiate between a spatial trend and a measurement uncertainty.

Project Logistics: Prior to the beginning of the design project, the instructor selects the student teams and their project managers. An attempt to equalize the capabilities of the teams is made by distributing the students on the basis of their academic standing, previous coursework, performance in the present course (including 4 problem sets and a midterm examination in the current course) and relevant work experience. The students self-assign other team jobtitles (e.g., project engineer, hydrogeologist) for their group. These positions are indistinguishable in terms of work execution, which is a group effort, and are essentially placeholders in ISIS cost-allocation algorithm. The exception is the project manager, who typically a graduate student is demonstrating a comfortable level understanding of the course material to date.

In the present course offering, the instructor-developed preliminary remedial assessment (Phase I) document for Camp Hazmat is delivered during a lecture in week 5 of a 10 week quarter. This document reflects the current scenario underlying the site thereby providing students with a general idea as to the main environmental concerns in the project. The instructor summarizes the scenario in a 30 minute oral presentation, which is followed by a 15 to 20 minute discussion of how to formulate the first work plan. Later that week, each team is guided through a one hour training session outside of normal class hours. The training session prepares the students for most of the basic ISIS tasks, including borehole drilling, soil sampling, well construction and groundwater sampling. More complicated features (e.g., sample analysis, field tests) are easily learned after the basics are mastered, and are not typically needed in the early phases of the project.

Preliminary Testing

A "beta" version of the ISIS package (ISIS v. 0.9) was used to deliver the 1999 environmental engineering design project. This project had been administered manually for each of the five previous years. ISIS 0.9 lacks a costing routine and several of the hydraulic tests discussed previously. The primary objective of the beta-test was to identify problems with the code. Significant problems are highlighted throughout the following sections. A second objective was to compare the ISIS-driven project with the manually administered version in terms of design project scope and timedemands imposed (on students and instructor). Rigorous assessment of this pedagogical strategy is available for the fully functional version of the program. However, a summary of student feedback emanating from the beta-test is provided at the end of this section.

Design Project Scope

In comparison with the manually administered version of the project, the ISIS hazardous waste site encompasses a substantially more realistic problem. First, the ISIS site encompasses an area about 35 times larger than the previous site. Second, the current version of ISIS can employ any number of contaminants; at present, it includes eighteen chemicals or chemical mixtures as opposed to five that were available in the manually delivered version. Finally, the introduction of the interactive hydraulic testing modules in the ISIS-driven project represents a substantial advance from the state of the manually driven project, where a hydraulic parameter was simply another datum provided to the groups upon request.

Regarding the simulator-driven PBL environment aspect of the project, the automated data acquisition aspects of ISIS greatly enhance project relevance over the manually administered project. The manually administered project often promoted student confusion and frustration due to the instructor based lag-time associated with all data dissemination tasks. For example, a group would often request contaminant samples at certain depth intervals, only to find upon data delivery that the soil lithology had changed in the latest soil-boring location, and that the requested depth-intervals were no longer optimal in terms of sampling locations. The ISIS-driven project empowers the students to make real-time adjustments in their work plans. This aspect of ISIS alleviates student frustration due to wasted labour, and thereby helps to maintain motivation during the course of a long-term design project.

Time Demands on Students and Instructors

The teams are scheduled for two, two hour sessions per week for five weeks (20 hours total) workstation sessions in a computer lab. Each team typically held meetings for a total of about two hours per week beyond this to delegate data-reduction tasks and to plan for the next computer session. Thus, all students were committed to about six hours per week outside of lectures. In the manually administered versions of the project, students held somewhat longer group meetings to assimilate hardcopies of fresh data, but had no computer sessions.

The number of hours dedicated weekly to the design project by the instructor, teaching assistant (TA) and programmer were tracked for each the past nine years. The design project was not offered prior to 1994, and the 1993 hours represent time dedicated to problem set administration and active office hours. Manual administration of the project by the instructor (1994-1995) imposed an excessive burden. In 1996, the project was also manually administered, but most of the workload was transferred to the TA. An initial investment in programming introduced (1996) afforded some relatively minor automated data acquisition features to assist the TA. Nevertheless, project administration demands remained excessive. The demands shown for the 1999 are inflated by programming problems encountered during first two weeks of the ISIS 0.9 test. The final values shown for the years 2000 and 2001 are based on subsequent courses taught using ISIS version 1.0.

Student feedback was gathered using a short questionnaire regarding the ISIS 0.9 driven project. The primary topic in the survey pertained to the intrinsic value of the project in helping the students to understand the course concepts. Other topics included were related to workstation accessibility, time demands of the project, and the usefulness of the virtual library. The student responses were uniformly positive as to the use of the ISIS project to supplement the course lecture material. Most of the students felt that the exercise helped them to achieve a deeper and longer lasting understanding of the subject matter. All of the students were enthusiastic about the convenience of the automated data distribution via email.

There were a variety of negative responses to certain aspects of the ISIS project. One of these pertained to the time-demands imposed by the project. Approximately one-half of the students felt the project load to be excessive, or felt pressured to complete the project in an unreasonable timeframe. Several students suggested that additional academic credit be assigned to what they considered to be a laboratory component of the course. Another common source of frustration for the students was the controlled access to the ISIS workstations. Many of the students would have preferred to concentrate their ISIS-workstation endeavours earlier in the design project period instead of distributing these throughout the project. The problems with the code in the early computer sessions clearly exacerbated this issue. Transporting ISIS to the personal computer platform has been successful and will help to alleviate the problem of access. In the more distant future, an Internet-based ISIS could increase access even more substantially.

Regarding the HTML-based virtual library, students indicated that they found it helped them to understand and select appropriate analyses and tests during their work sessions. The fact that they selflearned this material is corroborated by the fact that only a minimal amount of lecture time was dedicated to these analyses. Several students suggested that the virtual library be made Internet-accessible as well as ISIS-accessible to allow them review the material outside of the work sessions. This sentiment is consistent with the notion that the limited access to ISIS during the project was frustrating to many of the students. This chapter summarizes the development and features of the Interactive Site Investigation Software (ISIS) program, a PBL environment capable of administering an engineering design project. Key findings from a beta-test of ISIS in a senior-level design project are as follows:

1) The ISIS-delivered project tested here is broader in scope and can be more easily modified from one year to the next than the corresponding manually administered project. Thus, ISIS can greatly facilitate the infusion of recent research findings into the undergraduate curriculum.
2) The automated design project relieves instructors and TAs of tedious tasks (e.g., accepting work plans and disseminating data), while allowing the learner to work at his or her own pace.
3) This automation allowed learners to work more effectively, allowing the completion of substantially more virtual fieldwork than did the manually delivered project.
4) Student's responses to the beta-test were positive with respect to perceived usability and usefulness of the ISIS program. The most common negative response addressed the restricted access to the software.

The primary intent of using a program like ISIS is to automate the tedious aspects of a design project in order to free the instructor to participate more actively in terms of providing advice and feedback to the PBL-immersed students. The improved teaching and learning dynamics will presumably produce engineers with both a deeper understanding of design process and underlying principles.

4

Water Quality Management Systems

Full efficiency could be achieved by water quality control measures at the individual waste outfall or water supply intake was useful for discussion of the maldistribution of costs resulting from technological external diseconomies, and probably descriptive of a significant number of real cases as well. But engineering-economic research his shown that there are many instances where economies can be realised by collective measures such as joint waste treatment, lowflow augmentation for quality improvement, stream reaeration, groundwater recharge for quality improvement, effluent diversion or redistribution to make better use of natural assimilative capacity, the specialised use of streams or stretches of streams, and combinations of these. Where facilities involve scale economies that cannot be efficiently realised at individual outfalls or intakes, it may be appropriate to have a regional authority with powers more extensive than those assumed in earlier.

At one extreme would be a river basin or sub-basin agency empowered to plan, design, construct, operate, and finance virtually all water quality control measures; at the other would be an agency that simply sees to it through effluent charges or standards-that downstream costs are reasonably well reflected in the upstream decisions of managerially independent units. In this part we emphasize the opportunities which a regional authority might have to improve the efficiency of water quality management by implementing various measures, such as those mentioned above, that cannot be efficiently instituted at individual points of waste generation and discharge. However, we also recognise the importance of the technologic and economic linkages between water quality improvement as an output from water resources systems and other outputs from such systems, such as hydropower, navigation, and water-based recreation. For example, operation of hydro facilities for "peaking" can reduce water quality significantly by causing extreme low flows during "offpeak" periods. Water released from the deeper parts of storage reservoirs which have "stratified" may be devoid of dissolved oxygen and have an effect on water quality downstream from the reservoir similar to the discharge of an organic waste.

The installation of low head dams for slack-water navigation may result in substantial water quality deterioration. The dams reduce the velocity of flow and hold waste loads longer in a given reach of stream; at the same time they reduce the reaeration capacity in the given reach. Conversely, the operation of a storage reservoir to augment flows during summer periods to improve water quality may have adverse effects on water-based recreation on the reservoir, because of the resulting drawdown. These interrelationships suggest the desirability of a regional water resources management agency with responsibility for all aspects of water resources development and use. We return to this point where we discuss more specifically the question of appropriate institutional arrangements for water quality management.

Within the context of both water quality management and overall water resources management, a major problem is to integrate in an optimal way the decisions of fiscally independent units with those made directly by a regional agency. An agency concerned with water resources does not possess the scope, even in principle, to decide all courses of action bearing upon the way water resources are used. For example, if such an agency decided that a firm should be located at a particular site because the costs of handling its wastes would be low at that site, it might force the firm to incur transportation or other costs greater than the saving in waste management costs. Owing to such possibilities and even probabilities, decisions on location, production, and the character of production processes should *never* be based exclusively on considerations of water costs including waste disposal. Nevertheless, it is important to develop means whereby these considerations can be brought to bear on the water decisions of firms and of public agencies and on the decisions of agencies with broader responsibilities for land use planning. It is suggested in this section that systems of charges have special merit for this purpose.

This chapter focuses on the economic criteria for planning a regional water quality management system that includes collective facilities as well as those that can be implemented efficiently at individual points of waste discharge or water intake and use. The chapter also includes some discussion of system operation because planning and operation are interdependent. The specific components of an optimal system are often quite sensitive to the operating procedures adopted.

A regional agency has a variety of potentially efficient collective measures to choose from: increasing streamflows during low-flow periods via releases from reservoir or groundwater storage; treatment of intake water and of municipal and industrial wastes in regional collective treatment plants in compactly developed areas; artificial or induced oxygenation of watercourses (including reservoirs and lakes); diversion of effluents via pipeline or other means; specialised stream use; artificial recharge of wastewaters to groundwater aquifers for quality improvement; and the construction of shallow oxidation reservoirs in the stream itself. These methods could supplement measures to control the generation and disposal of wastes at individual points. A regional agency seeking to minimise the overall costs associated with waste disposal should plan and implement a system which equates the relevant incremental costs associated with water quality management in all directions. These costs include the costs of all manner of waste reduction and treatment and water supply treatment, the costs of flow regulation (including all opportunity costs that arise from the value of stored water for alternative uses), as well as the costs of conforming waste discharge to streamflow, changing industrial processes and outputs, and changing industrial location, and, finally, any damages which result from residual water quality deterioration. This statement that the agency should equate the "relevant" marginal costs does not mean that the marginal costs of waste reduction should be equal at all points of waste discharge. Indeed, in an optimal system, as explained in earlier, the costs of treatment or other waste reduction measures will tend to be higher at upstream points because upstream activities are reflected in downstream quality and costs. What the rule does mean is that it must not be possible to make marginal "trade-offs" and thereby reduce damage and waste reduction costs.

Ideally, the results of all relevant water resources systems and operating procedures would have to be considered and a solution derived which simultaneously indicated the optimum combination of systems elements and operating procedures. In industrial areas the solution would entail a system of charges and/or other measures such as effluent standards and zoning designed to secure the optimum amounts and locations of waste discharges. An "efficient" solution would enable the maximum net benefit to be obtained from the available water resources, and, as the latter implies, the overall cost associated with the disposal of an optimum amount of wastes would be minimised.

In economic terms "cost" means the value of foregone opportunities. In regard to water quality management such opportunities may be associated with resources used, for example, to build and operate treatment plants or reservoirs which could be used productively in other applications, or they may consist of opportunities to use water for purposes other than waste carriage or assimilation. The development in this chapter is simpler to visualise if the only output of the system is water quality control, so that all costs associated with waste disposal consist either of resources devoted to control facilities (reservoir regulation of streamflows, treatment of wastes, process changes, effluent diversion facilities, etc.) or of damages associated with water quality deterioration. The principles are exactly the same, however, if some of the costs consist of foregone benefits from other water-related outputs such as peak power.

The Basin-Wide Firm Again

The basin-wide firm is used here to illuminate economic principles. It is not put forth as a policy recommendation, for it would, in reality, have grave disadvantages: It cannot take account of uses that are not reflected in market prices; and, as a monopoly, it would present problems of regulation.

To illustrate some of the principles involved in taking account of the technical interdependencies in a river basin when economies of scale in water quality management measures are available, let us assume that a single firm (1) conducts all water-using industrial enterprises, all water and waste treatment facilities (no privately owned septic tanks, water softners, etc.), and all water transportation and related facilities; (2) operates all hydroelectric facilities, owns all land and structures in the flood plain, and is the sole provider of flow regulation; (3) controls all recreational uses of the watercourse; and (4) operates in competitive markets or in ones where public regulatory authorities set prices equal to marginal costs at levels of output that just clear the market.

In order to maximise its profits the firm would select the combination of water quality control measures (water supply treatment, waste generation and treatment, flow augmentation, wastewater releases coordinated with streamflow, etc.) and downstream damages that would minimise the overall costs associated with its waste disposal at its most profitable level of activity. This would be accomplished by equating the relevant incremental costs associated with waste disposal for all alternatives. As previously indicated, these costs would include any valuable outputs that are sacrificed to improve the efficiency of the waste disposal system.

If collective treatment costs are less than treatment at individual discharge points, the firm would transport effluent to collective treatment plants until the marginal costs of transportation and marginal treatment saved were equal. Since transporting wastes long distances and building scattered treatment plants are usually expensive procedures, the firm would weigh these and other diseconomies of scattered development against the economies (less congestion) to arrive at the optimum compactness of economic activities.

Compact economic development might be influenced by certain physical characteristics of the stream. Since the rate of reaeration through the air-water interface is directly proportional to the oxygen deficit, the organic waste degradation capacity is a decreasing function of the dissolved oxygen quality standards. This, plus the possibility of stream treatment to take advantage of economies of scale, emphasizes the possibilities of stream specialisation (i.e., using some stream or stretches of stream more heavily for waste disposal and maintaining higher quality in others) as a means of reducing the costs associated with waste disposal. Again the firm would trade off such possibilities against alternatives such as additional flow augmentation, higher levels of treatment, and other means of waste reduction. In deciding upon the location of a new industrial plant, the firm would consider not only transportation costs and other factors affecting assembly, production, and distribution costs but also the increment of total waste reduction and damage costs of alternative locations. The location producing lowest costs would be the one at which the marginal costs of the alternatives bearing upon overall costs are equalised.

While the point is expanded later, it is worth noting here that the firm "internalizes" all these costs. A water resources agency in a basin cannot internalize all of them, and *consequently it must attempt to have the opportunity costs of water resources use, including waste disposal, reflected in the decisions of entities that are simultaneously considering other resources costs.* This can be done by standards or, preferably, by charges.

The firm's general objective is to maximise profits, which implies that the costs associated with waste disposal will be minimised at the profit maximising level of output. If, for example, the firm could lower its costs by doing a little less waste treatment and permitting a little more damage, or by doing a little more water treatment and a little less waste reduction, or by a little more augmentation of low flows and a little less temporary storage of wastes, or a bit more process adjustment and a little less direct reaeration of the stream, etc., its overall profit position could not be at a maximum. As part of its general profit maximising activities, the firm would integrate its waste control activities with other aspects of its water-related operations. One major linkage is through flow regulation.

Thus, in computing the costs of alternative water quality control devices, the firm would need to consider complementary and competitive relationships among different water uses and alteration of the stream's flow regimen. Accordingly flow-augmentation costs would have to be determined in light of the fact that this alternative is often complementary with navigation, for example, but at least partly competitive with irrigation, flood damage reduction, power, and recreation. Any net benefits associated with other uses foregone in using flow augmentation to improve water quality in the optimum system are counted among the costs associated with water quality improvement. Moreover, the firm would consider the full marginal costs of producing particular products including the costs imposed on other activities by waste disposal. If the market adequately registered the population's evaluation of all goods and services sold by the hypothetical firm, its solution to the waste disposal problem would be "efficient," as that term is understood in economic welfare theory.

Many extreme assumptions have been made in developing this example; hence the basin-wide firm is not put forward as a realistic policy alternative. Still, the concept has provided a useful vehicle for illustrating applicable systems and decision criteria. The following points concerning public policy merit particular emphasis. First, public policies which explicitly recognise water quality management as a problem of optimising an interdependent system in a region will find many alternative control measures to consider. Accordingly, a least-cost system is more likely to result from the implementation of such policies than from the adoption of narrowly circumscribed con ventional approaches. The case studies in the next show the savings that can result from a more comprehensive and flexible approach.

Second, implementing the regional approach requires the planning, design, and integrated operation of many interdependent elements in the system. The illustrative firm "internalised" all these elements and, under our assumptions, found their optimum planning and operation in its interest as a matter of seeking maximum profit.

A water management agency should not attempt to internalize more elements than needed for its objective. In one instance, an agency may plan, design, and operate only collective measures such as reservoirs and reaeration devices and use direct controls or incentives (charges) to integrate these with the rest of the relevant economic system. In other instances a regional agency may find that it serves efficiency to go much further and it may plan, design, and operate virtually all water-related facilities in a region including water and wastewater treatment plants. Even in this case it would still be necessary to influence industrial process design and plant location. This could be done by direct regulation or, again we think preferably, by an appropriate system of charges.

In the following section we develop these thoughts further by using a somewhat simplified example which nevertheless permits us to address some matters of planning incentives and financing in a system which includes collective measures.

Regional System with Flow Regulation

We assume a system where the economically relevant alternatives are low-flow augmentation and waste reduction activities at individual points of waste discharge. The regional agency will design and operate a reservoir for flow regulation and use a system of effluent charges to control waste discharges at outfalls. If the reservoir is multipurpose we assume that the agency takes appropriate account of complementary and competitive uses of the water in determining the marginal cost of flow regulation for water quality management. For example, if navigation benefits occur in a fully complementary fashion with water quality benefits, the two must be summed and compared with the marginal costs of flow regulation.

Determination of the optimum amount of storage can be illustrated by a set of curves closely analogous to those used to illustrate optimum levels of waste reduction. Consider again a hypothetical streamflow-cost curve. Costs in this case are the total costs at various levels of flow of optimal combinations of downstream alternatives-for example, waste treatment, water supply treatment, and value of residual physical damages. We assume that these combinations are induced by an optimal system of effluent charges in accordance with the analysis presented.

The effect of flow regulation is determined as follows. Starting with a natural (unregulated) flow frequency curve, the effect of different levels of regulation on the flow regime is estimated. In this case, however, the downstream costs associated with a given flow remain the same, but flow regulation changes the frequency of occurrence of different flows. As before, the integrals of the curves are the expected values of the probability of costs. From a succession of such calculations a total downstream expected value of costs-avoided function can be constructed analogous, but in this case the damage reduction corresponds to increases in minimum flow rather than reduction in residual waste material.

A total cost function for the storage capacity necessary to produce the increases in minimum flow is also shown. Minimum expected total cost (or maximum expected net benefit if avoided downstream costs are termed benefits) is achieved at an increase in flows corresponding to point *X*. The slopes of the total storage cost and downstream cost functions are equal to their respective incremental functions. Accordingly at point *X* the optimality condition (i.e., equality of incremental costs) prevails.

The same solution can readily be stated in terms of functions which are more familiar from previous discussion. In general the optimality criterion is that the *expected value* of the incremental costs of all relevant alternatives should be equalised at the margin. If effluent charges are used to control individual waste discharges, the optimal condition requires that such charges reflect the incremental offsite costs imposed by each waste discharge. Charges will of course be lower when the assimilative capacity is increased by flow augmentation and/or by stream reaeration. In a correctly planned system the lowered charges reflect a gain in efficiency.

It may be worthwhile restating why point *X* represents an optimum combination of upstream and downstream measures. At this point a small further increase in reduction activity at the outfall would cost more than the expected value of downstream water treatment costs and residual damages avoided by downstream users. The argument is symmetrical for a slight decrease in reduction activity at the outfall.

When costs associated with waste disposal are minimised, the incremental costs associated with waste disposal are equated at each outfall. This means that the marginal costs associated with waste disposal are equalised in all directions. However, it does not mean that the marginal cost of reduction will be the same at each outfall along a stream. This would only be the case if downstream costs per unit of waste were equal for different outfalls, as they might be when two outfalls are across the stream from each other; then strict marginal cost equality would be necessary for cost minimisation.

Effluent Charges and Collective Measures

Fortunately, devising an appropriate effluent charges policy for an agency which incorporates economically desirable collective measures in its system requires no new principles whatsoever. The ideal charge will still be equal to the marginal external cost imposed. The residual external marginal cost imposed by waste discharge, when control at relevant points of waste discharge is induced to an optimal level by effluent charges, is the measure of the marginal benefit of the collective quality improvement facility say a flow-regulating reservoir.

Let us assume (unrealistically) that the agency creates the entire assimilative capacity of the river by producing a flow of water. Assume further that this flow can be produced at a constant marginal cost. Finally, assume that assimilative capacity can be increased efficiently in small increments. Under these assumptions, the agency, if acting efficiently, would increase assimilative capacity in such a way that the marginal cost of the last increment in capacity to control the effect of a waste would be equal to the effluent charge on that waste, which in turn is equal to the marginal cost of waste control at outfalls our "equal marginal cost" principle again. In this instance the proceeds from the effluent charge reflecting the marginal external cost would just cover the total cost of producing the assimilative capacity.

Given the same situation, except that we now assume that the marginal cost of further increments of capacity is rising, the relationship between the yield from charges and the total cost of the investment will change. The yield from effluent charges will now exceed the amount necessary to pay the total cost of the investment when optimum capacity is attained. Conversely, if the marginal cost of expanded assimilative capacity declines over the relevant range, the proceeds from the charge will be insufficient to cover the cost of the capacity expansion. This will be true even though the capacity expansion reduces the overall costs associated with the regional waste disposal system. If the regional agency decides not to augment assimilative capacity but to reduce waste loads by measures such as collective treatment, it should still follow the equal marginal cost rule and charge waste disposers the marginal reduction cost (which equals the marginal external cost) of handling the particular kind of waste involved. The assumption that the regional authority created all of the assimilative capacity of the stream makes for a theoretically neat outcome, but it is unrealistic because a stream always has some natural waste assimilative capacity. This circumstance in no way affects the use of the marginal cost rule as the appropriate criterion for the agency's investment decisions, but it does affect the financial results. When the agency levies a charge reflecting marginal external costs in the presence of natural assimilative capacity, part of the payment made is a "pure rent" on the scarcity value of the natural capacity. As we have indicated earlier, this is analogous to a charge levied on the use of public lands for grazing, for example. Accordingly, when natural assimilative capacity exists, the proceeds from the charge will be more than sufficient to pay for an economically justified augmentation of assimilative capacity produced under constant cost conditions. If an agency implemented only economically justifiable facilities and levied charges equal to the marginal external cost, it might take rather sharply decreasing costs in the provision of collective facilities for the agency to incur a deficit if there is a large amount of naturally occurring assimilative capacity initially available.

The above discussion has assumed that collective facilities can be expanded more or less continuously. Since some of them, such as reservoirs, involve substantial indivisibilities, this is not very realistic for small systems. But the principles are the same even when durable and lumpy investments are made. Of course it will be necessary to make long-run projections of population and industrial production and their related waste discharges with their attendant uncertainties. Other measures such as mechanical reaeration can be expanded more or less continuously. Groundwater recharge operations represent an intermediate situation. Moreover, as a system becomes larger and more diverse, even absolutely large additions become small relative to the total system and can be implemented on the basis of comparatively short-run considerations.

Unmeasured Values as Constraints on the Objective

If there were no problems in the manner in which the market registers water quality values, the policies described above would produce a solution quite consistent with the general rationale of a market system. But the market provides no value, or no satisfactory value, for some aspects of water quality, and this creates a special problem for public bodies because they must take account of all aspects. Considerable progress has been made in finding ways to assess the worth of such matters as general aesthetic effects, public health, and recreation, but dealing with these unmeasured values is still far from being a routine matter.

Two ways of handling unmeasured values come to mind. One way would be to label them "intangible" and disregard them in planning and designing water quality management systems. Then, when the proposed systems are presented for consideration and authorisation by representatives of the public, side information could be provided on aesthetic effects, public health, and other matters considered relevant to arriving at a decision in the public interest.

The second approach is to include hypotheses about such values in the process of system planning by expressing them in physical terms and treating them as constraints upon the cost-minimisation objective. For example, if social choice should dictate that the oxygen level in a stream must be high enough to support fish life at all times, the system must be planned to minimise the real cost associated with waste disposal subject to the constraint, or "minimum D.O. standard." This may require a different combination of units with different operating procedures than a system planned without the standard or constraint. If the constraint is effective and not automatically achieved when measurable costs (including damage costs) are minimised, it will increase the cost of the system, the extra cost representing the limitation which the constraint places upon the objective. To be consistent with efforts to achieve maximum welfare, the planning process should consider constraints "provisional" and view them as matters for research and study in order to discover how well they represent the preferences of society. One way of studying them from this point of view is to test their cost sensitivity. Varying a constraint by small amounts, redetermining the optimum system, and relating the change in costs to the associated physical changes in specific stretches of stream will provide information that will permit considered choices to be made by political representatives.

One useful way of stating the results of experiments with the constraints that are not valued directly by, or imputable from, the market is in terms of what they must "at least be worth." For example, a social judgment may be made that dissolved oxygen is to be maintained beyond the point indicated by the cost-minimising solution in order to preserve or enhance a recreational fishery. A comparison of the system with and without the constraint will not establish precisely what preserving the fishing pleasure is worth, but it will indicate the *least* value that must be attached to it if the higher level of control procedures is to be worthwhile. Or the additional cost of a higher level of certainty of achieving the desired dissolved oxygen level can be estimated. If constraints are imposed representing goals not directly commensurable with the values stated in monetary terms in the objective function, marginal conditions analogous to those indicated earlier must still hold if the cost-minimisation objective is to be fulfilled. The optimum system is not attained until a situation is reached in which it is impossible to make incremental "tradeoffs" between alternatives that will lower costs without violating the constraint. Once a combination of system elements has been decided upon, it will be necessary to utilise charges, effluent standards, or some other regulatory device to achieve optimum waste reduction at individual outfalls. The charge, if a charge is used, must be just high enough to achieve the standard, and it must not be possible to expand the large-scale measures at an incremental cost less than the incremental cost which can be avoided at the individual outfalls. If several wastes are involved, the expected value of all incremental costs avoided by an expansion in the large-scale measures must be added together and compared with the cost of expanding the measure.

We may generalise by saying that once a constraint is established by a duly constituted, well-informed political decision-making process, it may be viewed as representing a damage function which is perfectly inelastic at the specified quality level and degree of certainty. Large amounts of data and extensive knowledge of physical and economic relationships are needed to implement the procedure outlined above. Rapid and flexible computational techniques for estimating the physical characteristics and patterns of water quality and for carrying out the actual minimisation (maximisation) procedure are also essential. Some of these techniques, as well as the specific character of the information required to make reasonably dependable analyses of complex river basin systems, including cost sensitivity analysis of constraints, are illustrated by the studies described in the next.

Here we simply note that once a system is implemented the charges themselves are important sources of information about investment decisions in collective facilities. They indicate the marginal cost which waste disposers are incurring in order to meet the standard. This information can be used by the basin agency to determine whether a collective measure, either by capacity expansion or improved operation, can maintain the standard at a lesser incremental cost. Thus once the system is established the agency can obtain information on costs incurred by waste disposers without having detailed knowledge of their cost functions.

Institutional Restraints

Where there is no regional agency with authority to plan and operate a comprehensive system of quality management, the type of cost-sensitivity analysis outlined above can be used to illuminate the desirability of institutional change. In most basins in the United States the only collective measures are industrial and municipal waste-treatment facilities and reservoirs operated for low-flow augmentation by federal or state agencies. By comparing a least-cost system restricted to these measures with one that incorporates all relevant alternatives, a price tag can be put on the restriction. Or, to put it another way, the comparison will provide an estimate of the economic gain that society can derive from an institutional mechanism with authority to implement a wider range of alternatives.

Some "Short-Run" Aspects of System Operation and Cost Assessment

At this point it is useful to introduce once more the distinction between "long-run" or "planning" costs and "short-run" or "operating" costs. There are two reasons for making this distinction in the present context: First, a system should be operated on the basis of its "current" opportunity costs, especially when certain elements such as dams must be introduced in "chunks." Second, hydrologic or meteorologic fluctuation may make it desirable, in terms of cost minimisation, to operate the water quality control facilities in a variable manner. This indicates that, aside from any informational and administrative costs, effluent charges should be varied so as to maintain equality in the short-run marginal costs of alternatives.

So far the discussion has been primarily in terms of planning the regional system, and it was assumed that choices were not limited by past decisions. The view taken was sufficiently long-run, for example, to permit the building of dams *instead* of treatment plants. For planning, these are valid and appropriate concepts.

The point has been made several times that if the objective is to minimise the value of resources used internally and externally in connection with water quality management, the relevant cost for choosing a particular alternative is the value of the opportunities that must be foregone if it is chosen. This concept of "opportunity" costs has been used throughout, and methods of producing a better reflection of the value of foregone opportunities to society in decisions with respect to waste disposal were the subject.

The *concept* of opportunity costs does not change when a particular set of facilities has been decided upon and installed, but the *character* of the opportunities foregone does. When a dam is in the planning stage, all the labour, cement, steel, etc., needed for its construction can still be used in other activities, but once the labour, cement, etc., have been embodied in a dam they can no longer be used in alternative activities, and their price can no longer be considered an opportunity cost. The only costs that are still opportunity costs and thus relevant for further decisions are those that arise when a dam is operated for a particular purpose. If the dam is multipurpose, the opportunity costs may be largely "internal," representing the net value of the other uses foregone when its capacity is used for a particular purpose, although some operating costs will be involved, too.

In planning water quality management systems, therefore, the definition of costs should include all capital costs of new facilities. The system for water quality improvement should be expanded to the point where the cost of reducing an *additional unit* of any given waste or combination of wastes if the reduction activity involves joint products raises total reduction costs as much as it diminishes water quality damages. If there is a constraint, the system should be expanded until the constraint is met and the marginal total costs of all alternative measures for achieving the constraint are equalised. However, after a water quality management system is established, only the currently variable costs and opportunity costs internal to a multipurpose system are relevant. For example, assume that a decision is to be made about whether to retain a waste treatment plant with excess capacity or replace it with a smaller plant. Comparison of total costs, including capital costs of both plants, would almost certainly lead to the wrong conclusion. If the original plant has no alternative use, the relevant costs include only the operating costs, the land, and portions of the existing investment for which there are alternatives uses. The value of such uses must, of course, be brought into the comparison because they represent genuine opportunity costs. The test is always whether there are currently feasible, valuable alternatives. From the point of view of allocating resources to their most productive use, only inputs which have alternative uses can be said to involve costs.

Since storage reservoirs ordinarily cannot be economically constructed in such a way as to add small increments to system capacity, a system adjusting to changing waste load conditions will have periods of excess capacity. Under these circumstances the short-run opportunity costs of using the reservoir for flow augmentation for water quality improvement must be compared with the opportunity costs of alternative uses of the reservoir capacity. Appropriate operating procedures of the system would be based upon relevant marginal costs. Reliance upon flow augmentation would be heavier in the early period after the dam was built, and effluent charges or standards would be accordingly lower in those years. This means that care must be taken to provide waste disposers with information on longer-run prospects so that investment decisions will not be based on the presumption that charges will continue to be low. This chapter has focused on the integrated planning of various types of waste load reduction, flow regulation and in-stream measures; the incorporation of "unmeasured values"; and some "short-run" economic problems which stem from environmental variability. Attention was directed to the important role that can be played by effluent charges-either alone or in conjunction with other measures for regulating waste discharges in articulating the decisions of individual waste dischargers efficiently into a regional system for water quality management. In addition, it was pointed out that the assessment of charges is important in financing the collective features of such a system.

It was indicated that, ideally, system planning would consider all alternative uses of water, including its use for waste disposal, the effects of the various uses on water quality, the losses imposed on other uses by quality deterioration, and the value of water-derivative uses. All feasible system plans and operating procedures would be considered, and a solution would be derived which would indicate the optimum combination of system elements and operating procedures in light of the objectives and constraints relevant to the system. The regional system view, taking cognizance of the efficiency of collective measures, did not require the introduction of any new economic principles. But there are practical reasons, including the inherent complexity of the problem and the primitive state or absence of much of the required data, why an optimum solution cannot be fully attained. Furthermore, some inevitable arbitrariness in the specification of constraints, the absence of fully satisfactory means of handling the risk inherent in natural phenomena, and the uncertainty involved in forecasting future economic variables, militate against the achievement of fully optimum decisions. But with ingenious simplification, it is possible to develop workable procedures and mount investigations that will produce satisfactory, if not totally precise, results.

Nevertheless, an approach of the general type outlined in this chapter will improve decisions on water quality management. It emphasizes the importance of identifying interdependencies and viewing areas and economic functions tied together by "spillovers" in a "system" context; requires an explicit and unambiguous statement of objectives and constraints; stresses the identification and quantification of alternatives and their systematic testing in light of criteria of merit developed in terms of the objective or objectives of the system; and endeavours to cause the opportunity costs of water resources use to be reflected in decisions not controlled directly by the water resources management agency. These features of the analysis represent a way of thinking about the problem of water quality management; they do not imply a specific procedure for solving it. When the situation is a complex one involving numerous alternatives and constraints, optimisation procedures applied to formal mathematical models, often with the aid of computers, can be very helpful.

5

Hydrology and Hydrologic Function

Riparian areas link surface waters with the adjacent hillslopes. Riparian areas spread laterally across the floodplains onto the adjacent slopes of the watershed that contributes water to streams and rivers, and vertically from the top of the vegetation canopies to the groundwater aquifers. The character of vegetation, soils, channel morphology and ecological conditions within a riparian corridor influence the movement of water, sediment and nutrients within the riparian ecosystem. However, riparian areas in the arid and semiarid southwestern United States differ hydrologically from those areas in the humid eastern United States.

Variable precipitation and intermittent or ephemeral streamflows are more common in the Southwest than in the East. Baseflows are dependent on recharge by overland flow (surface runoff) generated by precipitation events, but baseflows in the East are dependent on groundwater flows and storm seepage. The Southwest has fewer lakes, ponds and other impoundments. Therefore, stream and river riparian systems are the focus of this chapter.

Precipitation Patterns

Precipitation patterns in the Southwest are influenced by the Pacific Ocean and the varying topography of high mountains, surrounding lowlands and desert floors. Annual precipitation ranges from more than 30 in. in the high-elevation conifer forests to less than 10 in. in the low-lying desert grasslands. Seasonal distribution of annual precipitation is also variable. A Mediterranean climate prevails in southern California and western Arizona, where most precipitation occurs during the winter when frontal storms move onto land from the Pacific Ocean. A bimodal precipitation pattern prevails in most of central Arizona. About 50% of the annual precipitation occurs as rain or snow during late November through late April. Another 35 to 40% of the precipitation is associated with localised, often intense, convectional rain storms during early July through the middle of September. Summer rainfall in the form of thunderstorms originating in the Gulf of Mexico provides most of the annual precipitation in New Mexico and western Texas. Occasional winter storms passing eastward from Arizona result in comparatively small but often widespread precipitation events.

Annual precipitation totals are subject to large temporal and spatial variability because of varying topography and erratic storm cell formation and movement. Therefore, average annual precipitation estimates are of little value in the management of water resources. Flows that are sustained by precipitation runoff are unreliable.

Soil Characteristics

Hydrologic processes in the Southwest are affected by the diverse physical, chemical and biological properties of soils in riparian corridors. Streams originating in high elevations flow through intermingling soils of volcanic, sedimentary or granitic origin, while low-elevation streams flow through alluvial deposits. Cycles of aggradation and degradation of sediments can restrain the development of well-defined soil profiles.

Soil moisture regimes can fluctuate from prolonged periods of soil saturation to conditions approaching aridity. Soil moisture is commonly less along low-elevation streambanks than those that are high-elevation. Evapotranspiration of riparian vegetation generally exceeds precipitation amounts because of the comparatively dry air and the high water table in many low-elevation floodplains. Coarse soils containing a high percentage of sands and gravels limit soil moisture capabilities and prevent shallow-rooted riparian plants from obtaining moisture when water tables decline.

Organic matter deposited onto riparian soils by leaf fall, sloughing of other plant parts and occasional flood events is less than in the eastern United States, and the decomposition of these accumulations is slower. Erodibility of soils is variable but is generally lower than in the East.

Limited localised salinity results from low precipitation combined with high rates of evapotranspiration. Excessive salinization can occur in internally drained basins but is rare in the absence of dams and other water impoundments. Low availability of nutrients in coarse soils constrains riparian plant establishment and growth, while the nutrient capital of fine-textured soils promotes better vegetative growth and development. Soil compaction and a loss of the litter layer also influence soil moisture and nutrient availability.

Runoff Processes and Streamflow

Only about 5 to 15% of the average annual precipitation in the region is converted into streamflow. Most precipitation is lost through evaporation and transpiration. Limited precipitation amounts infiltrate the soil surface and percolate downward through the soil profile to become subsurface flow to perennial streams or to recharge groundwater aquifers.

Management practices that increase the amount of available precipitation by reducing evapotranspiration from upland watersheds have long been of interest to the people of the Southwest.

Water Pathways to Stream Channels

Pathways of water flow to stream channels in the southwestern United States differ from those in the East, where subsurface flow moves vertically into the soil profile and then laterally into stream channels to become the major pathway of stormwater into and out of the riparian ecosystems of the well-drained forested watersheds. On the other hand, in the Southwest, overland flow plays a dominant role in moving excess water into stream channels.

Table: *Percent of Annual Precipitation Converted into Streamflow by Vegetation Type and Elevation Range*

Vegetative Type	*Elevation Range (ft)*	*Annual Precipitation (in.)*	*Streamflow (percent of annual precipitation)*
Mixed conifer forests	> 7,500	> 25	10-15
Ponderosa pine forests	5,500-8,000	18-30	10-15
Pinyon-juniper woodlands	4,500-7,500	12-20	10
Chaparral shrublands	3,500-5,500	15-25	8-10
Desert shrublands and grasslands	<3,500	<12	<5

Vegetation in the Southwest is sparse, soils are shallow and less permeable and high-intensity rain storms frequently exceed infiltration capacities of soils, leading to more overland flow. Rapid water-level rises recharge dry streambanks, resulting in large transmission losses, i.e., losses of streamflow to bank storage and channel bottoms when streamflow is first initiated in dry channels. Large bank storage and transmission losses reduce streamflows in intermittent or ephemeral streams.

Groundwater flows do not respond quickly to precipitation because of the long pathways involved. However, groundwater does contribute to streamflow in some instances. For example, groundwater discharge through springs provides the baseflow for the headwaters and a 25 mi reach of the upper Verde River in central Arizona. Sources of this baseflow were determined by analysing (1) historical records of water table levels, precipitation and streamflow; (2) downstream changes in baseflow; and (3) stable-isotope geochemistry of the groundwater, surface water and springs. The interconnected aquifers in this headwater area are the primary spring sources, supplying 80% of the baseflow. Increased groundwater pumping in the headwater area could lead to a dewatering of the river, consequently changing the character of the riparian corridor.

Factors Affecting Streamflow Response

The larger the watershed area, the greater the volumes and peak flows of streams and rivers. Round watersheds, as opposed to those that are more elongated, concentrate rainfall-induced streamflows more quickly at the outlet, so there is little difference between rainfall-induced streamflows and those from snowmelt runoff. Streamflow response to rainfall is quicker and peak flows are higher on high-elevation watersheds with steep hillslopes and channel gradients than on low-elevation watersheds with gentle hillslopes. Amounts and rates of overland flow from rainfall storms are higher on watersheds with soils of volcanic origin than on watersheds with sedimentary soils.

The greater the intensity and the longer the duration of rainfall, the higher the magnitude of the resulting stormflow. When a watershed has recently experienced a runoff-generating rainfall or snowmelt event, it is better "primed" to respond more quickly to subsequent events and produce a greater volume of streamflow water than a watershed with drier antecedent conditions. Precipitation and watershed characteristics combine to affect the magnitude and duration of stormflow response. Human activities can result in changes to watershed and stream channel conditions, which can affect streamflow response.

Impacts of Human Interventions

Large dams, reservoirs and water diversions built on the larger rivers for irrigation and municipal water use are major influences on water flows and hydrologic functioning of riparian corridors. Levees and floodwalls constructed along major waterways passing through urbanised centres to reduce the damage of flooding or other high-flow events can also increase the risk of flooding farther downstream because of a loss in the ability of streams and rivers to meander and, in doing so, reduce the energy of the flow. Increased velocities of water flow through levees and floodwalls can also change the downstream channel morphology and riparian ecosystem. Increased groundwater pumping for irrigation and municipal use has lowered some local water tables. Consequently, some of the perennial streams in the region have been transformed into intermittent or ephemeral streams. Dewatering of stream systems, at least seasonally, is also a possibility.

Flooding

Excessive flows in stream and river channels during and immediately after large high-intensity rainfall events has led to flooding, with associated economic losses and sometimes losses of human life. Flooding commonly results from excessive rainfalls associated with tropical storms from the Pacific Ocean and the Gulf of Mexico. Stream channels are altered by breached debris dams of uprooted trees and other accumulated materials, depositions of boulders within stream channels and streambank cutting and channel scouring. Streams can split into multiple channels or become braided, with water flows insufficient to support fish populations.

Future channel damage can result when unstable rock piles are moved by high streamflows following the flood. Unstable streambanks resulting from flooding are hazardous, particularly where trees and large boulders are precariously suspended high along vertical streambank faces.

Riparian vegetation often exerts a mitigating influence on flooding. Forest vegetation growing along headwater streams provides large woody materials that fall into the channels and reduces the damaging effects of rapidly flowing flood waters. However, tradeoffs must be considered in evaluating the role of riparian vegetation growing along floodplains.

Riparian vegetation can increase channel roughness and slow the velocity of flow, causing higher stages and more frequent overbank flows than unvegetated banks. On the other hand, the spreading out of water caused by riparian vegetation results in storage, which reduces downstream peak flows.

The distribution of many riparian tree and shrub species that are indigenous to the Southwest is dependent on periodic flooding to disperse seeds and create sandbars for seed germination. Restoration activities after flooding include actions to control or lessen the effects of boulder accumulations, timber and timber-related debris, stream channel diversions and channel scour. Stream stabilisation efforts that minimise sediment delivery into the channel help restore the capacity of a channel to transmit future floodwaters. Hydrologic and ecological functioning of many riparian areas have been restored by such efforts.

Erosion and Sediment Transport

Soil loss by water and wind action is greater in the southwestern United States than in the East. Much of the loss is due to surface erosion. Some gully erosion also occurs, but little soil is lost by mass movement. Upland erosion processes, streambank erosion and channel scour produce the sediments that are entrained in streamflows. Dynamic and complex fluvial processes and hydraulic factors are involved in the intermittent transport and deposition of sediments.

Soil Erosion

Most eroded soil particles are transported to stream channels during high-intensity rainfall events; little erosion is caused by snowmelt runoff in the southwestern United States. Soil particles from banks loughing and channel cutting are also deposited into channels. More soil erosion occurs on watersheds in the Southwest than in the eastern United States because the protective vegetative cover is more patchy and less dense and is, therefore, less effective in dissipating streamflow energy and stabilising streambanks.

Consequently, less water infiltrates into the soil and the greater surface runoff leads to increased erosion. Loss of the limited vegetation along riparian corridors and on hillslopes because of excessive livestock grazing, improper tree cutting or other human-induced or natural disturbances further accelerates erosion processes and the amount of sediment transported to the stream channels.

A limited amount of soil erosion and sedimentation can benefit riparian ecosystems. High streamflows or flooding can result in enhancement of soil development in low areas, increases in the fertility of soils forming the floodplain and replacement of the fine-textured soil particles needed for plant establishment and survival.

Sediment Transport

The episodic transport of sediments through stream channels mirrors the episodic streamflow patterns typical of the region. Snowmelt runoff events at high elevations are a major contributor to annual streamflow. A relatively slow response time from snowmelt runoff to peak streamflow is common, and associated flows in intermittent and ephemeral streams can last for several days or weeks. The energy available from snowmelt runoff events to move soil particles is relatively low and concentrations of sediments are low. However, much of the annual total production of sediments (suspended sediments and bedload) is a result of these events. The largest sediment loads transported in single hydrologic events are associated with the high-intensity, short-duration and mostly convectional rainfall occurring at all elevations during late summer to early fall. A rapid response time to peak streamflow discharge results and the streamflow duration is usually limited to hours or a few days. Intermediate amounts of sediments are moved as a result of the low-intensity, relatively long-duration frontal rainfall events in fall and winter. These latter events have a moderate response time to peak streamflow discharge.

Table: *Total Sediment (Suspended Sediments and Bedload) Transport after a 100 Year Storm by Vegetation Type and Management Practice*

Vegetation and Management	*Sediment Transport Total (lb/acre)*	*Suspended Sediment (percent)*	*Bedload (percent)*
	Ponderosa Pine Forests		
Thinning	1800	60	40
Clearing	17,600	90	10
Partial clearing and thinning	2400	90	10
	Pinyon-Juniper Woodlands		
Herbicidal conversion	115	80	20
Mechanical conversion	2200	73	27

Rain-on-snow events, while major streamflow-generators when they occur, represent less than 10% of the individual streamflow-generation events in the region. Sediment continues to be deposited in intermittent and ephemeral stream channels until a sufficiently large streamflow event occurs to move it farther downstream in pulses throughout the system. The factors responsible for the eventual sediment transport through a stream system are closely interrelated, complex in nature and difficult to quantify. Sediment movement in perennial streams is less episodic than in intermittent and ephemeral systems. However, flushes of sediment movement can occur in times of large rainfall or snowmelt events.

Suspended sediments represent the largest portion of the total annual sediment loads that move through stream systems in the region. As much as 90% of the total sediment transport is composed of suspended soil particles, with bedload particles representing a minor contribution. Similar percentages are also observed on an annual basis. However, the bedload component is more important in the structure and function of a stream. Downstream movement of bedload involves channel erosion and sediment deposition, which affects the stability of the channel.

Channel Dynamics and Morphology

Unlike in eastern watersheds, hillslope erosion in the Southwest is discontinuous and long lag periods between watershed events and sediment delivery often are common. The transport of sediment in streams and rivers is characterised as alternating pulses of aggradation and degradation that are punctuated by periods of inactivity. Macrotopography of the riparian corridors is rearranged during the alternating cycles of aggradation and degradation. These cycles often prevent the development of a stable organic soil horizon or a well developed soil profile, but, instead, they result in a soil that reflects the depositional patterns and history of channel flows.

Dynamic Equilibrium

A healthy riparian area maintains a dynamic equilibrium between the streamflow forces acting to produce change and the vegetative, geomorphic and structural resistance to this change. This relationship between sediment production and streamflow was originally discussed by Lane and later expanded by Heede, Rosgen and others to describe the changing nature of stream systems. When a riparian ecosystem is functioning in dynamic equilibrium, it is sufficiently stable that compensating internal adjustments occur without producing changes that overwhelm this equilibrium. This resilience or resistance to rapid change results from internal adjustments among factors such as vegetative structure, channel depth or stream morphology and other factors operating simultaneously to changes in the streamflow regimes or sediment movements. Excessive overland flows can increase channel flow volume and velocity, which causes channel erosion and increases downstream deposition rates. Under these conditions, the balance shifts back and forth and can be quickly moderated by internal adjustments so that no major change occurs in the dynamic equilibrium of the riparian area.

When the system's resilience to change has not been violated, a new equilibrium condition is established to sustain a healthy riparian area. Streamflows in excess of channel capacity overflow onto the floodplains, where healthy riparian vegetation and associated debris provide a substantial resistance to flow and act as traps for sediments. When floodwaters are dissipated in these areas, peak streamflow discharges are reduced and the deposition of coarse soil particles and other materials in the channel helps to trap the finer-textured sediments and accompanying nutrients. Bank overflows result in increased opportunities for germination and establishment of riparian plant species.

Channel Morphology

The morphology of a channel at a point in time is influenced by its width, depth and slope; the roughness of the contained materials; the stream velocity and discharge characteristics; and the sediment load and size of sediment particles. A change in one or more of these variables is likely to result in adjustments in the other variables and in the channel morphology itself. Contributing areas increase as one moves downstream, so the values of the variables also tend to increase and change the channel morphology.

Geomorphic and hydrologic variables have been incorporated into channel classification systems based on morphological characteristics. Rosgen's stream-channel classification is the most widely used system in the southwestern United States. This hierarchical system establishes discrete categories of channel types so that consistent descriptions and assessments of stream conditions are obtained for planning and management purposes. Stream-channel inventory information is organised into levels of resolution ranging from broad morphological characteristics (Level I) to more specific descriptions of stream reaches (Levels II, III and IV). While the procedure of classification relies largely on morphology, it is not readily apparent that the system is process based. Nevertheless, this channel classification system is grounded in the morphological process relations of fluvial systems.

In applying Rosgen's system, measurements of entrenchment, width-to-depth ratio and sinuosity are used to characterise a channel as one of eight major types. Single- and multiple-threat stream channels are also recognised. The major channel types are subdivided according to slope ranges and particle sizes of the dominant channel materials. These separations produce 41 initial classification possibilities. Rosgen's system can be applied to perennial, intermittent or ephemeral channels with little modification. Other observations and measurements (length of profiles, cross sections, etc.) necessary to morphologically classify stream channels by Rosgen's systems are presented in Rosgen and Brooks et al.

Classification by Rosgen's system of stream channels in the southwestern United States is incomplete; however, qualitatively speaking, headwater streams in the high elevations of the region tend to flow through stable, relatively straight, steep channels that are poorly to moderately entrenched, have small width-to-depth ratios and possess little channel sinuosity. Channels change little in their overall morphology once small streams meet to form tributaries to larger streams in the transitions of steep to level terrains at intermediate elevations. There can be increases in sinuosity, although the streams usually remain single-thread channels. Depending on the volume, velocity and timing of flows, the channels of large streams and rivers at low elevations are likely to be highly entrenched, have large width-to-channel ratios, be more sinuous and possess either single- or multiple-thread (braided) channels. Flood-plains of large streams and rivers in the Southwest are commonly flood prone.

Large Woody Materials

Tree and tree parts fall into stream channels because of streambank undercutting and collapse, tree blowdown or collapse and mass soil movements. Large woody materials that fall into a stream often become wedged in banks, increasing channel roughness and reducing streamflow velocities and retarding the downstream movement of large sediment particles when flows occur. Dissipating the energy of flowing water is an important function of fallen large woody materials. The larger the material, the longer it serves this function. Trees and tree parts extending into the channel deflect water currents laterally, causing the flow to widen the streambed. Sediments stored within piles of large woody materials add to the hydraulic complexity of a stream system, especially in organically rich stream channels that are wide and shallow with diverse riffles and pools in low-gradient of alluvial valley streams. Even if the stream becomes so large that downed trees cannot span the main channel, accumulations of large woody materials along the banks cause meander cutoffs and create well-developed secondary channel systems.

Water Quality

Riparian vegetation can reduce the amount of solar radiation impinging on the water surface, thus maintaining cooler water temperatures, which favour high-order aquatic organisms and maintain high dissolved-oxygen levels. This effect is a benefit in small headwater streams in the Southwest. However, removal of trees and tall shrubs along a streambank increases exposure of the stream to solar radiation and water-temperature increase. This leads to thermal pollution and deteriorated water quality in shallow low-volume streams.

Plant roots reduce the amount of bank cutting and erosion. This stabilising effect reduces sediment delivery to the channel and, therefore, helps to sustain the physical quality of the stream water. Riparian forests and other vegetative communities can improve water quality by removing or reducing sediments and nutrients in runoff water from upland agricultural or urban areas. Once in the stream, nutrient uptake by riparian vegetation transforms nutrients such as nitrogen and phosphorus from the soil solution and ties up nutrients in plant biomass. Riparian forests also improve water quality by creating more diverse stream habitats and favour organisms that more effectively remove nutrients and other pollutants from the water.

Figure : *A rosette sampler is used to collect samples in deep water, such as the Great Lakes or oceans, for water quality testing.*

Groundwater aquifers contributing to perennial streams contain dissolved constituents that affect the chemistry of the flowing surface water. The primary nutrients that are delivered to riparian areas by groundwater are nitrogen and phosphorus. Riparian ecosystems can reduce nitrate concentrations and, as a result, reduce groundwater nitrate concentrations and fluxes from the surrounding uplands to streams. Inorganic carbon can also be a significant dissolved constituent in areas where water passes through soils with high concentrations of calcite and dolomite. Return of treated wastewater or irrigation water containing fertilizers can detrimentally affect the quality of stream water by introducing unwanted chemical constituents into the systems. Some of these returns, however, have a value in recharging depleted groundwater aquifers.

Standards

In the setting of standards, agencies make political and technical/scientific decisions about how the water will be used. In the case of natural water bodies, they also make some reasonable estimate of pristine conditions. Different uses raise different concerns and therefore different standards are considered. Natural water bodies will vary in response to environmental conditions. Environmental scientists work to understand how these systems function, which in turn helps to identify the sources and fates of contaminants. Environmental lawyers and policymakers work to define legislation with the intention that water is maintained at an appropriate quality for its identified use.

The vast majority of surface water on the planet is neither potable nor toxic. This remains true even if seawater in the oceans (which is too salty to drink) is not counted. Another general perception of water quality is that of a simple property that tells whether water is polluted or not. In fact, water quality is a complex subject, in part because water is a complex medium intrinsically tied to the ecology of the Earth. Industrial and commercial activities (e.g. manufacturing, mining, construction, transport) are a major cause of water pollution as are runoff from agricultural areas, urban runoff and discharge of treated and untreated sewage.

Categories

The parameters for water quality are determined by the intended use. Work in the area of water quality tends to be focused on water that is treated for human consumption, industrial use, or in the environment.

Human Consumption

Contaminants that may be in untreated water include microorganisms such as viruses and bacteria; inorganic contaminants such as salts and metals; organic chemical contaminants from industrial processes and petroleum use; pesticides and herbicides; and radioactive contaminants. Water quality depends on the local geology and ecosystem, as well as human uses such as sewage dispersion, industrial pollution, use of water bodies as a heat sink, and overuse (which may lower the level of the water).

The United States Environmental Protection Agency (EPA) limits the amounts of certain contaminants in tap water provided by US public water systems. The Safe Drinking Water Act authorizes EPA to issue two types of standards: primary standards regulate substances that potentially affect human health, and secondary standards prescribe aesthetic qualities, those that affect taste, odor, or appearance. The U.S. Food and Drug Administration (FDA) regulations establish limits for contaminants in bottled water that must provide the same protection for public health. Drinking water, including bottled water, may reasonably be expected to contain at least small amounts of some contaminants. The presence of these contaminants does not necessarily indicate that the water poses a health risk.

Some people use water purification technology to remove contaminants from the municipal water supply they get in their homes, or from local pumps or bodies of water. Water drawn directly from a stream, lake, or aquifer and has no treatment will be of uncertain quality

Industrial and Domestic Use

Dissolved minerals may affect suitability of water for a range of industrial and domestic purposes. The most familiar of these is probably the presence of ions of calcium and magnesium which interfere with the cleaning action of soap, and can form hard sulfate and soft carbonate deposits in water heaters or boilers. Hard water may be softened to remove these ions. The softening process often substitutes sodium cations. Hard water may be preferable to soft water for human consumption, since health problems have been associated with excess sodium and with calcium and magnesium deficiencies. Softening may sacrifice nutrition for cleaning effectiveness.

Riparian Buffer Strips

The maintenance of riparian buffers of trees, shrubs or herbaceous plants is an effective way to reduce the velocity of surface runoff that can carry sediments and associated pollutants from upland areas; these pollutants are trapped in the litter within the buffer area. This filtering action prevents some pollutants from being deposited in the stream channel. By reducing sedimentation, the chemicals attached to sediments (phosphorus, heavy metals, etc.) are prevented from entering water bodies. Shade cast by trees in the buffer strips also helps to reduce thermal pollution in adjacent water bodies.

In contrast to the management of riparian buffer strips in the eastern United States, emphasis in the Southwest is placed less on planting vegetation and more on managing the naturally regenerated riparian vegetation. However, the re-establishment and subsequent protection of herbaceous plant cover along severely degraded streambanks is practiced occasionally.

6

Perturbations of Hydrological Cycle

The provision of water-supply, sanitation and drainage are key elements of the urbanisation process. Substantial differences in development sequence exist between higher-income areas, where the process is normally planned in advance, and lower-income areas, where informal settlements are progressively consolidated into urban areas, but common factors are impermeabilisation of a significant proportion of the land surface and major importation of water from outside the urban limits. Sanitation and drainage arrangements are also fundamental to consideration of the urban hydrological cycle. They generally evolve with time and vary with differing patterns of urban development, but installation of mains sewerage generally lags considerably behind population growth and water-supply provision.

Urbanisation causes radical changes in the frequency and rate of subsurface infiltration, with a general tendency for volume to increase significantly and for quality to deteriorate substantially. These changes cannot be measured directly, and are thus difficult to quantify, but in turn influence groundwater levels and flow regimes in underlying aquifers. Subsequently groundwater quality degradation, both within the urban area itself and in downstream alluvial aquifers, occurs.

Quality Deterioration within Urban Limits

Some urbanisation processes cause radical changes in the quality of subsurface infiltration. This is widely the cause of marked, but essentially diffuse, pollution of groundwater by nitrogen compounds, increasing salinity, elevated dissolved organic carbon concentrations (which can lead to enhanced mobilisation of Fe and/or Mn) and, on a more localised basis, contamination by faecal pathogens and/or micro-organic pollutants. The intensity of impact varies widely with the pollution vulnerability of underlying aquifers and with the type and stage of urban development. In alluvial formations the uppermost unit in the commonly present multi-aquifer sequence is vulnerable to pollution from human activities at the land surface, given its shallow water table.

Rapid urbanisation and industrialisation with indiscriminate use of the ground for liquid effluent and solid waste disposal presents a complex array of activities potentially polluting to groundwater. In many districts without mains sewerage, a heavy subsoil contaminant load originates from in-situ sanitation and the disposal of sullage waters increases the risk of shallow groundwater contamination, because of the presence of various household chemicals. In addition to elevated nitrogen concentrations, increased concentrations of chloride (mostly from excreta), sulphate and borate (from detergents) and bicarbonate (from oxidation of organic matter) are frequently observed. A further, increasingly-frequent, cause of shallow groundwater contamination in residential areas of developing cities is hydrocarbon fuel leakage from underground storage tanks at gasoline filling stations.

In many rapidly developing cities an increasing number of industries (such as textile mills, tanneries, metal processing, vehicle maintenance, laundry and dry cleaning establishments, printing and photoprocessing, etc) are located in the extensive fringe urban areas without mains sewerage. Most of these industries generate liquid effluents, such as spent lubricants, solvents and disinfectants, which are often discharged directly to the ground and can represent a long-term threat to groundwater quality. Santa Cruz-Bolivia is a low-rise, relatively low density, fast growing city, whose municipal water supply is derived entirely from wellfields within the city limits, extracting from a semi-confined, outwash-plain, alluvial aquifer. The city has relatively high coverage of mains water-supply, but only the older central area has mains sewerage, and most domestic/industrial effluents and stormwater drainage are disposed to the ground. The uppermost aquifer unit thus shows substantial deterioration in groundwater quality down to depths of about 40 m. Groundwater abstraction from the deep alluvial aquifer has induced downward movement from the shallow horizons and a component of contaminated water is now observed at depths approaching 90m. The heavy development of the shallow aquifer for private water supplies, however, effectively provides a degree of protection for deeper municipal wellfields by abstracting and recycling part of the polluted water, which is fortuitously good management practice provided that none of the supplies provided by these wells are destined for sensitive use.

In some cities located on low-lying coastal alluvium, direct disposal of wastewater to the ground via onsite sanitation is not possible and effluents are discharged into rivers and canals, which often become influent to aquifers as a result of groundwater abstraction. Hat Yai-Thailand is an example of this condition. Most of its limited mains water-supply is imported from external surface water sources, but about 60% of the overall supply is provided from local groundwater resources. The disposal of domestic and industrial effluents to the ground by onsite sanitation systems is not always possible because of the shallow water table and such wastes are discharged into rivers and canals. The heavy groundwater abstraction results in induced infiltration to shallow aquifers detectable most readily by high NH4 -N concentrations, reflecting the low Eh of the aquifer system. With increasingly heavy groundwater development it is believed that induced canal seepage is now the largest component of groundwater recharge.

Downstream Riparian Effects

Although the provision of mains sewerage lags considerably behind population growth and water-supply provision, sewage effluent (termed here wastewater) is generated in large volumes by the majority of (but not all) rapidly-developing cities. This wastewater is normally discharged to surface watercourses after minimal treatment from where, especially in more arid climates, it is used on an uncontrolled basis for agricultural irrigation in downstream riparian areas. Such areas are usually underlain by important alluvial aquifers and examples of this situation include many cities in northern Mexico and northeastern China.

Infiltration of wastewater either through agricultural soils or via stream bed recharge can result in groundwater quality degradation including elevated chloride, nitrate or ammonium and some phenols. In Leon-Mexico, the groundwater quality of deep municipal boreholes within an area irrigated by wastewater, is being threatened by increasing salinity with downward movement of a Cl front.

Consequences of Uncontrolled Aquifer Exploitation

Groundwater quality issues cannot be divorced from those of resource exploitation. The most common quality impact of inadequately-controlled aquifer exploitation, particularly in coastal situations, is the intrusion of saline water. For thin alluvial aquifers this takes the classical wedge-shaped form, but in thicker multi-aquifer sequences salinity inversions often occur with intrusion of modern seawater (or retention of palaeo-saline water) in near surface aquifer horizons with fresh groundwater below.

Contamination of deeper (semi-confined) aquifers, where they underlie a shallow poor quality phreatic aquifer affected by anthropogenic pollution and/or saline intrusion, is a frequent consequence of uncontrolled exploitation. This occurs as a result of inadequate well construction leading to direct vertical seepage and/or pump-induced vertical leakage, with penetration of more mobile and persistent contaminant species. Evidence has been accumulating since the 1980s of widespread drawdown of the piezometric surface by 20-50 m or more in various Asian megacities, as a result of heavy exploitation of alluvial aquifers, and both of the aforementioned side-effects are quite widely observed.

A recent Asian Development Bank technical cooperation programme on water resources management in megacities included case histories for 4 Asian cities in the humid tropics, which possess major alluvial groundwater resources. The results of these studies have been reviewed, and amplified by further direct data collection and other references, with the aim of drawing generic conclusions. Among the cities surveyed, groundwater remains the major component of municipal (public) water supply only in Dhaka-Bangladesh, having been substituted in other cases by long-distance imports of surface water. This was often due to quality deterioration through saline intrusion and/or anthropogenic pollution, but sometimes the result of reduction of individual borehole yields, due to falling water-table or poor well construction and maintenance.

The situation is not as simple as it might at first appear, however, since in the other cases (Bangkok, Jakarta and Manila) resultant shortage and increasing cost of water-supplies led to a major growth in private well drilling, such that the overall exploitation of groundwater increased, despite attempts to initiate control, as a result of fears about further saline intrusion and/or land subsidence. There is little point in controlling municipal abstraction if private groundwater exploitation is not similarly managed. In effect, what has occurred in Bangkok, Jakarta and Manila is the replacement of a moderate number of municipal groundwater supplies, which were at least capable of being systematically controlled, monitored, protected and treated, by a very large number of shallower, largely uncontrolled, unmonitored and untreated sources.

Tapping groundwater at location of demand makes sense for many industrial users and for amenity irrigation, since these are difficult to supply through mains distribution. However, it is questionable in densely-populated areas, both on economic and on public health grounds, as regards domestic supplies and those for sensitive industries (such as food and beverage preparation). An added concern is illegal connection of private wells to the mains water-supply, without measures to prevent 'back syphoning' at times of reduced mains pressure, with consequent contamination of 'down-system' users.

Implications for Groundwater Management

Rapid urbanisation has been shown to have a profound effect on groundwater recharge, flow and quality. The scale of implications for the security and safety of developing city water-supplies is considerable. The paper has selected examples from tropical alluvial aquifers in view of their importance internationally, and brevity prevented presentation of a wider spectrum of hydrogeological conditions. Nor is it possible to discuss systematically all the resource implications and the corresponding management strategies.

While, in many instances, institutional and regulatory arrangements will need strengthening to implement management strategies, such strategies must also be soundly based on hydrogeological criteria if they are to achieve more sustainable exploitation of groundwater resources. Some of the ways in which hydrogeological considerations can be meshed into a groundwater resource management programmes are defined below.

Aquifer Pollution Control

Given typical socioeconomic and hydrogeological conditions, it is not realistic to attempt to protect shallow alluvial aquifers from some quality deterioration during the urbanisation process. However, it will be prudent to control those activities which most threaten groundwater quality overall, especially that in deeper (less vulnerable) aquifers. This will be best achieved through the following strategy:

- undertake a rapid survey of subsoil contaminant loading to identify those activities likely to pose the greatest threat to groundwater through mode or intensity of discharge and presence of persistent and/or toxic chemicals,

- establish the degree of existing deterioration of the uppermost aquifer unit from anthropogenic pollution by a sampling survey of representative shallow wells, with analysis for appropriate indicator determinands,
- introduce selective controls on subsoil contaminant loading (where demonstrated necessary) through extension of mains sewerage, incentives for improved handling/control of industrial chemical effluents, etc.

Resource Management Criteria

The use of urban groundwater also needs to be directed and rationalised, taking account of the quality distributions and trends identified. A better balance needs to be achieved between shallow and deep groundwater abstraction, for example by directing nonsensitive water users towards exploitation of groundwater of inferior quality so as to minimise the downward migration of high concentrations of anthropogenic contaminants.

Municipal groundwater development strategies need to be harmonised with private groundwater use patterns and with wastewater disposal and/or reuse strategies. The hydrostratigraphical and geohydrological complexity of tropical alluvial aquifers means that groundwater recharge evaluation is always subject to considerable uncertainty, and rarely forms a sound basis for resource management.

Other simple criteria are needed and combinations of the following approaches, appropriately adapted to local hydrogeological conditions, are likely to prove more useful:

- determine the extent of vertical layering of the aquifer system, and associated hydraulic head and groundwater quality variations, by selective field data collection,
- appraise the susceptibility of the system to saline intrusion and/or land subsidence following major depression of the piezometric surface, in a qualitative sense on simple geohydrological criteria,
- use target piezometric levels as the resource management criterion, since this is more likely to maximise groundwater production while minimising irreversible side-effects which threaten sustainability,
- spread public water-supply abstraction geographically to avoid large cones of piezometric depression in the deeper semi-

confined aquifers, especially in areas susceptible to saline intrusion and/or land subsidence,

- in situations where significant encroachment (or intrusion) of saline water has already occurred, avoid abandonment of pumping from salinised wells and try to encourage continued abstraction for appropriate uses to reduce landward hydraulic gradients,
- avoid generation of large downward hydraulic gradients by balancing the abstraction between shallow and deep aquifers, by encouraging nonsensitive users (industrial process and cooling water, amenity irrigation, etc) to drill shallow wells and by reserving deeper aquifers for potable supplies and sensitive industrial uses; this may be achieved by direct licensing controls or differential abstraction tariffs.

Direct and Secondary Water Benefits

The studies surveyed in this chapter provide estimates of direct and total benefits that may be expected to accrue from the use of water for agricultural purposes in various western states. Some of the studies measured only the direct benefits; others extended the analysis to include secondary benefits to market-related activities. While any program or project analysis should attempt to encompass total benefits, the conditions under which economically legitimate secondary benefits will exist are sufficiently complex and ill-understood that it seemed desirable to report separately on the more straightforward analyses of direct benefit estimates before getting involved in the measurement of secondary benefits.

None of the studies summarised here deal specifically with large-scale interbasin transfers of water. Their relevance stems from the fact that they measure benefits from agricultural applications of water, and it is just such applications that would have to absorb most of the water provided by the proposed large-scale transfers. Agriculture currently accounts for almost 90 per cent of consumptive water use in the West and constitutes the only potential demand that is large enough to absorb the supplies of proposed transfers (from 2.4 to 110 million acre-feet per year). Further, the water supply problems of agriculture are less susceptible to solution through technological change and innovation than are the water supply problems of municipal and industrial users whose basic consumptive requirements are much lower.

Benefits in Groundwater Areas

The price charged for water can be taken as an upper bound on direct benefits to the immediate user when supplies are sufficient to permit the full adjustment of the quantity applied so that it results in maximum profits to the water-using activity. This adjustment can almost always take place in irrigation areas supplied primarily by groundwater. Since most state laws in no way restrict pumping, the usual "common pool" rationale (if I don't get it, my neighbour will) leads pumpers to apply water until the net value of its marginal product equals the unit cost of pumping.

Evidence on the value of water at the margin of application is thus provided by data on pumping costs. Taylor (1964) quotes pumping costs in the Madera Irrigation District (California) in 1961-62 averaging $5 per acre-foot with a variable component of $1.96. Young and Martin state that variable pumping costs in Central Arizona range from $7 per acre-foot with a 315-foot lift to $12 with a 540 foot lift. In the Texas High Plains, Grubb (January 1966) shows pumping and delivery costs of $8.74 per acre-foot. The evidence from these major potential importing regions points to water-use values having been pushed fairly low at the margin in terms of direct benefits.

Brown and McGuire (1967) studied the problem of optimising the distribution of surface water among the constituent districts of the Kern County Water Agency (KCWA) and simultaneously optimising the pumping of groundwater. KCWA had contracted with the State of California for ultimate delivery of 1.1 million acre-feet, and the problem was how to price the water to guarantee that it would be fully utilised by the districts and that the allocation among districts would be economically efficient. A major component of the study was the estimation of the water demand function for each district, i.e., the marginal value of water in each district as a function of the total amount applied.

Functions were fitted to two sets of data, one from the California Department of Water Resources and the other from a farm budget study for the districts in KCWA. The subsequent optimisation analysis to two optimum prices to be charged by KCWA to its constituent districts, depending upon which demand function was accepted as the more accurate. The cost of delivering water from KCWA to the district and the average delivery cost within each district were added to the optimum prices to arrive at the delivered prices (prospective marginal values of water).

Table: *Prospective Marginal Values of Water in the Irrigation Districts of KCWA under Optimum Allocation*

		Delivered price (= marginal value) per acre-foot	
District	***Annual 1, 000 acre-feet***	***Case 1 delivery***	***Case 2***
Belridge W.S.D.	206	$18.25	$14.60
Lost Hills W.D.	168	14.50	10.85
Rosedale-Rio Bravo W.S.D.	70	19.20	15.55
Semitropic W.S.D.	127	17.70	14.05
West Kern County W.D.	3	22.50	18.85
Antelope Plain W.D.	59	23.20	19.55
Wheeler Ridge-Maricopa W.S.D.			
No. 1	142	23.30	19.65
No. 2	29	28.75	25.10
Kern River Delta and others	309	18.00	14.35

Source: Brown and McGuire (1967).

When the districts served by KCWA begin receiving the water supplies already contracted for from the State Water Project, the averaged marginal value of water in that large area will be approximately $19 per acre-foot (Case 1) or $15 (Case 2), depending on which demand function is assumed. Additional water would yield direct benefits no higher than those figures. Moore and Hedges (1963), whose detailed farm budget studies formed the basis for the Case 1 demand function mentioned above, derived demand functions for different sizes of farms representing a cross section of sizes found in Tulare County, California. In doing so, they also took account of differences in soil types, and in the acreages that could be used for high-value crops (cantaloupes, sugarbeets, and cotton).

The initial application, marking the transition from dry farming to some irrigated acreage, produces a rather high value per acre-foot, ranging from $77 to $82 per acre-foot, depending on farm size. Thereafter, the marginal value falls off steadily with increasing applications. It is precisely this phenomenon of diminishing returns plus downward pressure on market prices which limits the amount of water that can be used efficiently in agriculture.

Hartman and Anderson (1962) studied the value of supplemental irrigation water in Northeastern Colorado through an analysis of farm sales data from the North Poudre Irrigation Company over the period from 1954 to 1960.

This company started taking delivery of Colorado-Big Thompson (C-BT) water through the Northern Colorado Water Conservancy District in 1957. The company has issued shares which entitle the holder to a specified percentage of the total supply available in a given year. When farms are sold, the shares of stock are generally transferred with the farm, although there is an established market for these shares within the company. The analysis, consisting of a regression of farm sales price on assessed value of buildings, acres of farmland, and shares of irrigation company stock transferred, indicated that a share was worth $105 in 1954-56 (before GBT deliveries) and $198 in 1960 (after GBT deliveries had begun). These figures are in general agreement with some independent sales of shares at about $70 in the early 1950s and at about $200 in the early 1960s. Water deliveries per share averaged 3.3 acre-feet in 1954-56 and 6.4 acre-feet for the 1957-60 period. Thus the average capitalised value per acre-foot of annual delivery for the increment of C-BT water was about $30 ([$198-$105]/[6.4-3.3]). *The annual value* for the incremental water is approximately $3 per acre-foot if a 10 percent capitalisation rate is assumed, or $1.50 if a 5 percent rate is assumed.

Anderson (1961) studied the irrigation water *rental* market in the South Platte Basin. Rental prices for water on an interim basis are particularly suitable measures of the marginal value of water in a region such as the South Platte Basin where water rights are owned by irrigation companies and are readily transferable on a temporary basis within and between companies. Since there are many potential buyers of temporary rights to excess water, competition is likely to force the rental price up to the marginal value in the most productive use. There may, however, be exceptions to this stemming from community pressures to maintain a customary price. Table below indicates rental prices charged by representative irrigation companies in 1959.

Table: *Rental Prices of Water, South Platte Basin, 1959*

Company	***Rental price per acre-foot***	
	Early season	***Late season***
North Poudre Irrigation Company	$2.50	$4.20
New Cache La Poudre Company	3.25	3.25
Greeley and Loveland Company	3.00	5.00
Water Supply and Storage Company	5.00	8.00
Farmers' Reservoir and Irrigation Company	4.60	6.00
Larimer and Weld Irrigation Company and Windsor Reservoir Company	2.70	2.70

Source: Anderson (1961).

Crop irrigation accounts for more than 90 percent of the water *consumed* annually in Arizona. Of the water consumptively used on crop irrigation, about 60 percent is used to irrigate extensive crops such as alfalfa hay, sorghum, and barley and 40 percent is devoted to high-value intensive crops such as cotton, vegetables, and fruits.

Young and Martin (1967) studied the value of water in Arizona agriculture by constructing budget studies for a farm typical of Central Arizona. The characteristics of the farm were synthesised from surveys of over 600 farms.

These figures indicate the *maximum* amount that such a farm could pay for water from any source for use on particular crops, covering only variable costs, and leaving no return for coverage of taxes, insurance, depreciation, interest on investment, or management. At water costs in excess of $34 per acre-foot (the highest "short-run ability to pay" figure for cotton), it would no longer pay this typical farm to produce cotton even if other crops were sufficiently profitable to pay all the farm's overhead. If water costs exceeded

Table: *Typical Water Use and Average Short-Run "Ability to Pay" for Delivered Surface Water in Arizona*

	Pumping lift		
Crop and item	***315 feet***	***460 feet***	***540 feet***
Upland cotton:			
Water use per acre (acre-feet)	6.0	5.0	5.0
Income over variable costs per acre ($)	134	117	110
Income over variable costs per acre-foot ($)	22	23	22
Variable pumping costs per acre-foot ($)	7	10	12
Average short-run ability to pay per acre-foot	29	33	34
Barley:			
Water use per acre (acre-feet)	3.0	2.5	2.0
Income over variable costs per acre ($)	32	22	18
Income over variable costs per acre-foot ($)	11	9	9
Variable pumping costs per acre-foot)	7	10	12
Average short-run ability to pay per acre-foot	18	19	21
Alfalfa hay:			
Water use per acre (acre-feet)	6.1	6.1	6.1
Income over variable costs per acre ($)	33	13	5
Income over variable costs per acre-foot ($)	5	2	1
Variable pumping costs per acre-foot ($)	7	10	12
Average short-run ability to pay per acre-foot	12	12	13

Contd...

	Pumping lift		
Crop and item	**315 feet**	**460 feet**	**540 feet**
Grain sorghum:			
Water use per acre-foot (acre-feet)	3.3	2.75	2.2
Income over variable costs per acre ($)	33	23	18
Income over variable costs per acre-foot ($)	10	8	8
Variable pumping costs per acre-foot ($)	7	10	12
Average short-run ability to pay per acre-foot ($)	17	18	20

Source: Adapted from Young and Martin (1967).

Note: Variable pumping costs are added to "income over variable costs" to arrive at "ability to pay" because the availability of delivered surface water would eliminate the need to pump.

$13, alfalfa hay would be dropped. At $20, grain sorghum would be dropped; and, at $21, barley would be dropped. In the longer term, the ability to pay for water would be reduced by an additional overhead cost of at least $8 per acre-foot that would have to be covered if this typical farm is to remain in business.

It seems safe to conclude that the marginal value of water lies below $21 per acre-foot for the production of barley, alfalfa, and grain sorghum the types of agriculture which account for more than 50 percent of total consumption in Arizona. Goss and Young (1967) compiled data on the pricing policies of the major water distributing agencies in Central Arizona. The usual caveat applies to using prices as a surrogate for marginal values (i.e., direct benefits).

Stults (1966) studied the impact on agriculture of the falling groundwater table in Pinal County in south-central Arizona. This county accounts for 25 percent of the state's cropped area, and includes 38 percent of the cotton planting.

The study analysed the changes that are likely to take place between 1966 and 2006 as a result of the falling water table (increased pumping costs), on the assumptions that optimum adaptation of cropping patterns took place, that the margin of returns over nonpumping variable costs remained the same, and that technology and government programs (including cotton allotments) remained as in 1966. The results were as follows: (1) a 50 percent drop in total acreage was projected from 259,000 acres to 129,000 acres; (2) a 42 percent decline in water use amounting to 415,000 acre-feet; and (3) a 20 percent drop in net farm income amounting to $5.4 million per year. These results imply a value in terms of income forgone averaging $13 per acre-foot by the year 2006.

Grubb (January 1966 and October 1966) studied the value of irrigation water to the Texas High Plains. Since the late 1940s, there has been a shift from dry farming to irrigated agriculture based on groundwater pumping, and the irrigation of more than 5 million acres is causing the water table to fall by 3 to 4 feet per year.

The direct benefits per acre-foot for a delivered surface supply replacing an exhausted groundwater supply consist of the difference between irrigated and dry farming incomes plus the pumping costs avoided.

For a "composite acre" representative of the entire High Plains agriculture, Grubb projects direct benefits to 2020 under the following assumptions:

(1) the composite acre, consisting of 32 percent cotton, 38 percent grain sorghum, and 21 percent wheat, remains constant;

(2) crop yields per irrigated acre remain constant, but irrigation efficiency improves so that application per acre declines from 13 inches to 9 inches between 1970 and 1990;

(3) prices of products and inputs remain constant; and

(4) in the absence of irrigation, dry farming would be practiced with the same percentage distribution of crops.

The average benefit figures provide an upper limit on the marginal direct benefits of water. Thus it appears that marginal direct benefits for the High Plains are currently no more than $27, with this limit increasing gradually to about $36 in the future as irrigation efficiency increases.

More detailed evidence is provided by Grubb (October 1966) on returns to irrigation in the Southern High Plains by crop and soil type. From those data, estimates of marginal benefits have been derived for different levels of water application. If cropping patterns are similar to those used by Grubb in his "composite acre" and if a preplant application plus 3 postplant applications is the general practice, are generally in agreement with the $27 to $36 average benefit.

Table: *Projected Benefits from a Delivered Surface Water Supply, Texas High Plains*

Year	***Direct irrigation benefits peracre-foot***[a]	***Ground water pumping costsper acre-foot***[b]	***Average surface water benefits per acre-foot***
1959	$18.20	$8.74	$26.94
1970	18.20	8.74	26.94
1980	26.59	8.74	35.33
1990-2020	26.80	8.74	35.54

a. Calculated from Grubb (January 1966), Report No. 11, using the assumptionof 1.1 feet per acre through 1970 and 0.75 feet per acre thereafter.

b. It has been assumed that the fuel and irrigation equipment items of Grubb's represent present and prospective costs of pumping and delivering groundwater. Part of these costs could not be escaped even if the water source were other than groundwater.

Richard L. Johnson (1966) has made a study of the marginal value of irrigation water in the 9,000-acre Milford, Utah, area. The methods used were to estimate a Cobb-Douglas production function for agriculture in that area and to construct a linear programming model of the area's agriculture. Both of these methods produce direct estimates of marginal water benefits. Part of the intent of the study was to compare the estimates produced by the two methods.

Table: *Estimates of the Marginal Value of Water in the Milford, Utah, Area, Derived from Three Production Models*

	Marginal Value		
Acre-feet of water used per acre	***Cobb-Douglas model***	***Linear program I***	***Linear program II***
1	$22	$14	$42
2	14	14	15
3	11	14	15
4	9	14	14
5	8	0	6
6	8		

Source: Richard L. Johnson (1996)

Most farms in the area applied between 3 and 5 feet of water per acre, and for these rates of application, the marginal value of water implied by the analysis ranged from zero to $15 per acre-foot, depending upon the production model used. Fullerton (1965) and Gardner and Fullerton (1968) studied the water rental market in the Delta area of Utah, an area of about 180 square miles in the west-central part of the state. The total cultivated area varies from 35 to 60 percent of the total in different years, and 80 to 90 percent of this is in alfalfa. Four irrigation companies serve the area, diverting and storing water from the Sevier River. Prior to 1948, only *intracompany* temporary transfers among water users were permitted. Since 1950, exchanges on an intercompany basis among all companies have also been permitted. Naturally, it would be expected that the efficiency of water use would be improved by the greater flexibility of transfer. In the 1965 study it was found that the mean rental price was $3.21 per acre-foot up to 1948 and $9.60 after intercompany transfers began. Corrected for changes in price levels and for changes in the efficiency of water delivery, the latter price becomes $8.75, a measure of the marginal value of water under the more efficient water marketing arrangements. In the 1968 study the average price from 1955 to 1964 was reported to be approximately $12 per acre-foot.

Direct Benefits and Return Flow Data

The studies surveyed above indicate that the value of direct benefits to the agricultural user of water at the margin of application varies from place to place but is generally low. The figures revealed by the studies are listed to indicate the range of values, but it should be kept in mind that some of these figures are *upper bounds* on direct benefits and may overstate actual direct benefits.

The figures revealed only a range of benefits to direct water users and do not include an allowance for values generated through return flows. Return flows from agriculture are generally no more than 40 percent of withdrawals, but for illustration, let us assume 50 percent consumption and 50 percent return flow. If time lags are not important (as in a region of year-round irrigation), the value of the return-flow multiplier might be as high as 2. At the other extreme is the case where return flows have no effective value because they move into aquifers from which they cannot be recovered a condition that apparently holds for Central Arizona where percolation rates are less than the rate of fall of the water table. However, if the growing season is less than a year, time lags in return flows can be quite important.

For example, suppose that return flows are one-half of withdrawal and that 50 percent of the return takes place within one month, 80 percent within two months, and 100 percent within three months.

Table: *Summary of Direct Benefits per Acre-Foot at the Margin of Application: Western United States*

State	***Benefits per acre-foot***
California: KCWA (after State Water Plan)	$ 19
Colorado:	
N. Poudre Irrigation Co. (rental)	3
S. Platte Basin	3-8
Arizona:	
Central Arizona, short-run	21
Central Arizona, long-run	13
Major distributing agencies, Central Arizona	0-10
Pinal County	9
Texas:	
High Plains, average, now	27
High Plains, 1990	36
Southern High Plains, hardland soils	10
Southern High Plains, mixed soils	45
Utah:	
Milford area	0-15
Delta	12

If a four-month irrigating season is assumed, the actual multiplier value would be 1.34 rather than the limiting value of 2, largely because late irrigations contribute only to offseason return flows. However, as off-season return flows could be of value to other water users (e.g., for electric generation or navigation) and could be stored and used for irrigation in a later season, a reasonable upper bound on the return flow multiplier for most regions might lie in the neighbourhood of 1.5. For areas other than the groundwater dependent areas of Texas and Central Arizona, an assumed return flow multiplier value of 1.5 would raise the direct benefits as high as $30 per acre-foot in the surveyed parts of California. For the groundwater dependent areas of Texas and Central Arizona where values to the initial user are highest, a return flow multiplier much in excess of 1.0 would seem questionable. One can therefore conclude that benefits per acre-foot at the margin of application to direct water users in agriculture, including the value of return flows, generally lie below $30 per acre-foot in the major water-using areas covered by this survey.

The Secondary Benefits Issue

To arrive at an estimate of total benefits, it is necessary to consider secondary benefits as well as direct benefits to water users. The direct users of water are tied to other sectors of the regional and national economy; and, as their activities expand or contract, so will the activities of those they supply or from whom they buy. The extent of these induced secondary changes and their relevance to an assessment of national benefits and costs depend very much on the circumstances of the industries and regions involved, as well as on national economic conditions. The basic issues in determining the presence and magnitude of any legitimate secondary benefits are:

1. Which industries will contract as well as expand, and where are they located?
2. Over what period do labour and capital immobilities exist and how are they finally resolved?
3. What happens in the long run to the productivity of displaced factors?
4. What economies of scale exist in expanding and contracting
5. What economics and diseconomies of scale are involved in providing public overhead services and housing for labour which migrates as a result of project construction and operation?

Without answers to these questions, one cannot know whether there are legitimate secondary benefits, let alone what their magnitudes might be. This is a much neglected field in economics and in public policy formulation. However, there have been several attempts to measure total benefits per acre-foot used in agriculture, and some of these are reported below.

Hartman and Seastone (1966) analysed the income losses stemming from the diversion of water from agriculture to municipal and industrial uses in northeastern Colorado. The gross value of products produced per acre-foot of water applied was found to be $27. Because there is a significant return flow to surface sources, the authors used a return-flow multiplier of 2, making their estimate of the total value of agricultural output $54 per acre-foot. The direct profit and rent components in agriculture and in its supplying industries were estimated to be $14.50 per acre-foot, of which $12.90 was directly in agriculture. Thus, if there were no problems of labour immobility (no loss of labour income), the income losses would be somewhere in the range of $12.90 to $14.50 per acre-foot, the exact value depending upon the opportunities for the suppliers to pick up equally profitable business. This is what water would be worth to farms over the life of their present equipment and land improvements.

The total wage and salary income related to gross production of $54 was estimated at $17.20 ($10.25 for agriculture and $6.95 for related industries). If labour were immobile in both agriculture and the related sectors, $17.20 would have to be added to the opportunity cost of transferred water over the period of immobility. Thus, the benefits forgone from water transferred from agriculture to municipal and industrial uses would range from $14.50 to $31.70, depending upon the degree of labour mobility. The periods of labour immobility applicable to the various sectors are not known.

Hartman and Seastone (1970) also considered potential income losses stemming from the loss of agricultural water rights in the Imperial Valley. Their analysis utilised a California input-output model of Martin and Carter (1962) and resource requirements matrices of Zusman and Hoch (1965) to trace the indirect effects of the reductions in Imperial Valley production. Not only were the usual "backward linkages" of input-output analysis taken into account, but possible impacts on "forward linked" industries supplied by Imperial Valley agriculture were also analysed. Such forward linkages can be important only when no close substitute commodities are available at comparable costs to sustain the dependent industry. Hartman and Seastone concentrated on the particular forward linkage of food and feed grains to cattle feeding and meat processing. Their analysis showed total direct and indirect income losses of $32 per acrefoot of water used in the feed grain sector, but a total of $106 per acrefoot if a strict forward linkage to cattle feeding and meat processing is assumed. The authors conclude that $18 of the $106 represents income completely unrecoverable because of long-term immobilities. Thus, for a short period and assuming a rigid forward linkage, income losses might be as high as $106 per acre-foot. In the longer term this would drop to $18 as capital is recovered and as labour mobility increases.

Grubb (January 1966) studied the incomes generated directly and indirectly by irrigated agriculture in the Texas High Plains. As reported earlier in this chapter, he found direct benefits in agriculture to range from $18 per acre-foot presently to $27 per acre-foot after 1980. In addition, he estimated value added in the processing and marketing of irrigation output, in the provision of agricultural inputs, and in the satisfying of consumer demands stemming from those additions to income. To interpret Grubb's figures, one must relate the notion of value added in related industries to the appropriately defined measure of secondary benefits. Value added equals the total income payments made to capital, land, and labour as the result of some productive process.

Grubb found direct benefits plus value added in the related industries to range from $81 per acre-foot to $119 per acre-foot after 1980. These figures substantially overstate *national* economic benefits per acrefoot of water supplied, because of the implicit assumption that all indirect inputs would be unemployed in the absence of irrigated agriculture. More realistically, when importation of replacement water is considered in order to sustain irrigated agriculture, at least some of the inputs would have alternative employments outside of agriculture, so that only part of the indirect value added can legitimately be counted as secondary benefits and only over the period when the inputs would have remained unemployed. The proportion of value added and the period would have to be determined by a much more thorough investigation of the local economy.

Furthermore, as pointed out by Professor Kelso (personal correspondence dated 19 July 1968), Grubb's indirect values added were not reduced by the amount of capital and operating inputs imported into the region. Thus the above figures would overstate benefits even under conditions of complete immobility of resources. Wollman et al. (1962) studied the value of water in alternative uses in t San Juan and Rio Grande basins in connection with the diversion of water from the San Juan River (a Colorado tributary lying mostly in New Mexico) into the Rio Grande Basin via Rio Chama. This area serves as a microcosm for the entire semiarid, rapidly growing Southwest because of its urban-oriented growth and the signs of a growing disparity between the supply of and demand for water. Mining, agriculture, and industry are the major water users in descending order of total value of output in the region, but agriculture accounts for about 99 percent of total consumptive use.

New Mexico was awarded 838,000 acre-feet per year from the San Juan River under the Upper Colorado Interstate Compact. Wollman's objective was to determine the effects on the state's economy of different patterns of use by agriculture, industry, mining, and recreation of the 638,000 acre-feet then remaining unappropriated. Eight possible patterns of use were evaluated four patterns of allocation under each of two schemes which called for the diversion of 110,000 acre-feet and 235,000 acre-feet per year. Total value added, total employment generated, and total water-related costs were estimated for each of the eight patterns of water use.

An additional comment on the use of value added as a measure of benefits is required before proceeding. It was argued that value added directly and indirectly represents an *upper bound to shortterm benefits* accruing to water provided in rescue operations. The New Mexico case, however, is one of a region expanding on the basis of a new water supply. What relation does value added bear to national benefits in such a situation?

A growing region that wants to stimulate economic activity and attract labour and capital from the rest of the economy might use the total value added (direct and indirect) in the region per acre-foot of water as a criterion for the selection of water uses *provided* the inputs used in expansion are not simply bid away from other activities in the same region.

Even under these circumstances, the total value added per acre-foot is not necessarily correlated with an industry's ability to expand profitably. For example, assume that a region has 100 units of water to allocate between two industries and that the value added per unit of water consumed has been $100 for industry A and $1,000 for industry B. Making the 100 units available to industry B may seem to imply a potential regional income ten times that which would result from allocating all of the water to industry A. However, this does not of necessity follow from the conditions stated. An industry can be high in value added without being a high profit industry, and it is profitability that is the signal for expansion. Limited markets and high costs for inputs and transport may severely limit the ability of industry B to expand. Thus, value added is of limited usefulness as a valid regional criterion for the allocation of new water supplies to expanding activities without studies of product markets and input availabilities.

The relationship of value added in an expanding activity to national economic benefits is not, in general, at all clear except that value added per acre-foot constitutes an upper bound on national benefits.

Wollman et al. found that value added directly and through purchases ranged from \$28 to \$51 per acre-foot in agriculture, from \$212 to \$307 per acre-foot in recreational use, and to much higher figures in municipal and industrial uses.

This again establishes a fairly low upper bound on benefits per acre-foot in agriculture in the San Juan and Rio Grande Valleys of New Mexico.

Estimates of the value added directly and indirectly per acre-foot of water intake for different agricultural sectors of the Arizona economy are shown. These estimates were derived by adding water costs to the data on personal income (wages, rents, profits, and interest) derived by Young and Martin.

So long as water is used for the production of food and feed grain, no new supplies can be credited with benefits in excess of \$28 because the benefits of a proposed project cannot exceed the cost of providing the same service by the best alternative means. Imported water cannot be assigned benefits in excess of the incomes that would be lost if an equivalent amount of water were removed from existing agriculture.

Under the conditions found in rescue operations, short-term total benefits may somewhat exceed \$100 per acre-foot on the basis of evidence gathered in this chapter.

These high values would obtain for only short periods after the introduction of new surface supplies, however, and would diminish to values not exceeding \$30 to \$50 per acre-foot. In situations characterised by the expansion of new areas, as in the New Mexico and Arizona cases, the upper bound on total benefits ranges from about \$30 to \$50.

To illustrate the importance of the period of immobility of resources, let us consider a hypothetical situation using the Texas High Plains figures. It is assumed for illustrative purposes that complete capital and labour immobility in agriculture and related industries obtains from the time water is withdrawn or exhausted until year *T*, at which time all inputs except those still engaged directly and indirectly in dry farming would be able to move into new activities. The benefits from replacement water will be taken to be \$119 per acre-foot from year zero to year *T* and then \$27 per acre-foot from year *T* until year 50 (the assumed life of a water transfer project).

Table: *Total Benefits per Acre-Foot in Agriculture*

State of region	***Direct benefits***	***Total benefits***
Northeastern Colorado	\$12.90	\$14,50-31.70
California, Imperial Valley (rescue operation)	-	106 very short term
	18 long term	
Texas High Plains (rescue operation)	18.20-45.00	81-119 short term only
New Mexico	-	28-51
Arizona	9-21	28

Note: These figures are from the regional studies reported in this chapter; they are based on differing assumptions and should not be compared with one another without an understanding of these assumptions.

For different periods of immobility, the average annual value per acre-foot of water computed at 5 percent interest over the 50 year life of a water import project would be as follows: \$36 ($T = 2$); \$49 ($T = 5$); \$66 ($T = 10$); and \$90 ($T = 20$). This simple analysis makes it clear that the period of factor immobility is a crucial factor in the determination of benefits from rescue-type water transfers.

A caveat should be added before concluding this discussion. Even though total benefit figures can under certain circumstances be rather large for short periods, the ability of the direct users to *pay* for water will generally be much smaller because nearly all of the components of value added that were counted in total short-run benefits under factor immobility represent out-of-pocket costs to the farmer. Thus the financing of rescue operations will probably have to involve not only charges to direct users for water delivered but also taxes on the wider set of community economic values sustained by the water importation.

The Future Demand for Output From Irrigated Areas

The preceding sections have indicated ways of assessing the direct and total benefits from irrigation water at the margin of application. Such benefits are, of course, ultimately derived from the demand for the crops produced in irrigated areas, and a growth in demand for irrigation output could change the benefits picture. Future demands will depend on the world food situation and on the federal government's future policies with regard to the role of the United States as a supplier of world food.

Only a few years ago, it was universally true that rates of population growth exceeded the growth of agricultural production in the underdeveloped areas. World grain stocks had decreased sharply from 1960 through 1966. In 1966, nearly one-fourth of the U.S. wheat crop was shipped to India alone.

By 1967 signs of change were being perceived. A detailed projection study by Abel and Rojko (1967) indicated that by 1970, world grain production and consumption would likely be in balance. The study further indicated that if the underdeveloped countries continued to make moderate improvements in productivity and if the 1967 acreage of 165 million or more were to remain under crops in the United States, the world would continue to have excess production capacity through the 1980s. The improvements in agricultural productivity which have taken place in the less developed countries, particularly in Southeast Asia, have exceeded all the expectations of 1967. The exciting story of the "green revolution" and its implications for world grain production has been told by Lester R. Brown (1970).

If the underdeveloped nations are able to handle the tremendous social and economic changes imposed by the startling spread of highly productive grain culture, and if the aid-giving nations continue to assist in building up the infrastructure necessary to gain full advantage from the production breakthrough, Brown predicts not only self-sufficiency in grains for a large part of the underdeveloped world but possibly a reversal of export flows toward some of the more developed countries. The United States has excess agricultural capacity that can be utilised on relatively short notice. About 56 million acres has been withdrawn under various government programs. Upchurch stated in 1966 that in a period of a few years, a large part of this idle acreage could be brought back into production at about present levels of prices. If limitations on the acreage in cotton are continued and, at the same time, restraints on production of food and feed grains are relaxed, the United States could supply an additional 60 million tons of grain. Under present conditions, there seems to be no indication that anywhere near this much will be needed for food aid in the near future.

The Conservation Needs Survey of a few years ago showed that the United States had 638 million acres of land in the three classes best suited for cropping and that this land was available without further drainage or irrigation. In recent years only about 200 million acres have actually been used. This land availability plus continuing innovation in plant breeding and food technology indicate a vast, untapped potential for increased production within the confines of presently croppable lands in the United States, Any likely demands of the next several decades, including the U.S. share of world food aid, appear quite capable of being met without opening up new lands.

7

Estimating Hydrogeological Conditions

For a number of years United Kingdom Nirex Limited (Nirex) investigated the area around Sellafield in NW England to assess its suitability to host a deep repository for solid intermediate-level and some low-level radioactive wastes. These wastes originate from Britain's nuclear programme and largely derive from the reprocessing of power station fuel. The investigation involved the drilling of 18 boreholes to depths of up to 2000 m and the collection of extensive sets of hydrogeological and hydrochemical information from down-hole tests. The data supported numerical models of the groundwater flow system and these were used to investigate whether dissolved radionuclides from a potential repository could be transported back to the surface in significant amounts. Initial scoping calculations showed that the flow systems and particularly the groundwater salinity distribution predicted by these models respond slowly (> 100 ka) to changes in driving conditions.

This response time is long compared with the duration of the present climatic conditions of the Sellafield area (of the order of 10 ka). The area was glaciated several times during the Quaternary, most recently within the last 20 ka, and the changes caused by glacial processes could still be affecting the hydrogeological system. The lifetimes of some of the wastes considered for disposal exceed the c. 100 ka cycle of glacial-interglacial episodes over the latter part of the Quaternary, therefore the potential impact of further glaciations has to be considered in assessments of the long-term safety performance of a repository.

Although the Nirex investigation of the Sellafield area is now concluded, the issue of glacially affected hydrogeology is relevant to a radioactive waste repository located anywhere in the UK, or in similar areas relative to glacial maximum ice sheets, even in areas outside those expected to be covered by the ice sheet.

This chapter is a case study that explores the difficulties in producing a description of the effect of major climate change on a groundwater system, for future incorporation into groundwater modelling calculations of the long-term safety performance of a potential repository. The description needs to address the response through glaciation and associated absolute sealevel change, but this is complicated by isostatic changes in land elevation and by erosion and deposition. The description must be consistent across these. The possible effects are explored for the sellafield area in the context of global climate changes over the last 120 ka, broadly encompassing the whole of the last interglacial-glacial-interglacial cycle, i.e. the Ipswichian-Devensian-Holocene (Recent) in British terminology. This work has been reported in full by Nirex (1999). Evidence of the actual behaviour of the system is considered, primarily based on groundwater chemistry and the nature of the youngest minerals infilling pores and fractures.

Quaternary Hydrogeological Conditions Relevant to Britain

Hiscock & Lloyd (1992) quantitatively examined the hydrogeological behaviour of a shallow Chalk aquifer in eastern England over the Devensian, using a vertical section, numerical model that incorporated density-driven flow associated with salinity. The model was first calibrated to approximate present-day observations of salinity distribution, using present-day boundary conditions of sea level and position, precipitation and recharge. The model was then extended into the past using analogue climate data from Goodess et al. (1988) to specify recharge under previous climates. Spelcothem age data from Gordon et al. (1989) were used to estimate likely periods of groundwater recharge, in that speleothcms are unlikely to form at times when groundwater recharge is inhibited by permafrost (perennially frozen ground). Sealevel boundary conditions at various past times were estimated. No explicit account was taken of land surface erosion, or of movements of land and sea elevation.

Groundwater flow regimes were calculated for the Ipswichian interglacial (at c. 124ka BP), the early Devensian cold period (at 100 ka BP), and for two intervals (10 ka and 7.5 ka BP) after the main late Devensian glaciation, the Dimlington stadial glaciation. No estimate of conditions was made for the Dimlington stadial, although the presence of an extensive till of this age attests to the presence of ice. The main evidence used in validation of the model was the evolution to the present-day distribution of water chemistry in the Chalk. Important findings were the significance of permafrost in greatly reducing recharge during cold climates, and the development of increased transmissivity of the Chalk aquifer in the post-glacial period.

In a more general study, van Weert et al. (1997) presented a large-scale groundwater model of the southern North Sea basin for six climate scenarios from present-day temperate conditions to the most extensive ice cover corresponding to the Anglian (Elsterian) glaciation. The hydrogeology of the bedrock for the area was simplified to two aquifers separated by a low-permeability layer, all of this overlying impermeable basement. Where appropriate, the top boundary conditions for the bedrock formed the bottom boundary conditions of an ice-sheet model.

During the glacial maximum, large quantities of subglacially derived meltwater were predicted to have discharged through thick sand aquifers in some areas; in other areas the aquifer was unable to transmit the water, and tunnel valleys to convey the water were predicted to form at the ice-rock interface. Because of the scale of the model, considerable simplification of details was necessary. Piotrowski (1997) discussed van Weert et al.'s modelling and emphasised the importance of subglacial drainage where bedrock permeability is low. More recently, Boulton et al. (200 k) have reviewed the possible effect of long-term climate changes, including glaciation, on a deep geological repository in Sweden. The results present estimates of the thermal, hydrological and mechanical effects of a regional ice sheet, determined from a simplified ice-sheet model. A particular conclusion is that groundwater chemistry may be affected to depths of up to 2000 m.

Setting of the Sellafield Area

This study of the Sellafield area parallels that of Hiscock & Lloyd (1992), but treats the phenomena in greater detail, taking into account a greater body of information and recent developments in understanding of cold-climate hydrogeology. The Sellafield area is situated on the NW coast of England, adjacent to the Irish Sea. The coastal lowlands (average elevation <100m above present mean sea level (pmsl)) are underlain by Triassic sandstones up to 1000m thick, with a locally important aquifer in the more permeable upper part of the sandstones. Offshore to the west in the East Irish Sea, the sandstones are overlain by a thick sequence of Triassic mudstones and evaporites, the Mercia Mudstone Group. To the cast, the land rises steeply towards the Cumbrian mountains of the Lake District, reaching elevations of over 800 m above pmsl within 15km of the present-day coast.

The mountains are formed of Ordovician to Silurian igneous and low-grade metamorphic rocks of low permeability and are deeply dissected by glacially overdeepened valleys. The Triassic sandstones of the coastal lowlands and the Triassic mudstones of the East Irish Sea are overlain by a varied sequence of Quaternary deposits, mainly dating from the last (Devensian) glaciation. There is evidence of a main glacial advance over the area during the late Devensian followed by a retrcat-deglaciation and at least two glacial readvances. The ice had a far-travelled 'Scottish' or a local 'Lake District' origin, depending upon location. The boundary between the ice masses varied with time but the 'Scottish' ice never penetrated much inland of the present coastal zone. With local exceptions, such as the deposits underlying the Late Devensian till at Drigg, pre-late Devensian Quaternary deposits from the onshore area appear to have been largely removed. The geological setting has been described in more detail by Michie (1996) and Akhurst et al. (1997). The hydrogeology of the Quaternary deposits has been described by McMillan el al. (2000). An account of the groundwater chemistry has been given by Bath el al. (1996). Fresh groundwater is ubiquitous at shallow depth and is separated by a transition zone at -400 to -600 m above Ordnance Datum (OD) from the deeper saline waters across the entire region. In the west, the saline waters are Na-Cl-dominated brines derived from solution of Triassic halite in the Irish Sea Basin. In the east, the saline waters are more dilute and have a different chemistry (although still dominated by Na-Cl) derived from water-rock interaction beneath the Cumbrian Mountains.

The history of the last glacial-interglacial cycle in west Cumbria is better understood than preceding cycles, and an outline chronology from the Ipswichian onwards is given. The absence of unambiguously dated horizons from within the onshore Late Devensian sequence precludes establishment of a precise chronology for the area. After the interglacial warm phase, there was a long cold phase in the early Devensian with only local ice-sheet development, although understanding of the extent and timing is still developing. The growth and coalescence of an extensive ice sheet over most of the British Isles, finally centred on the western Highlands of Scotland, seems to have developed from about 25 ka BP, but it may have been growing as early as c. 29 ka BP. Most workers place the period of maximum glacial extent, the peak of the main Late Devensian Glaciation, equivalent to the global Last Glacial Maximum, at 18-20ka BP, although some have dated it earlier at c. 22 ka BP. Lamb & Ballantyne (1998) argued that the ice in the Lake District reached an elevation as high as 800-870 m above pmsl, with isolated nunataks protruding.

Deglaciation left the Sellafield region ice free no later than c. 14 ka BP. Readvances of regional (Scottish) ice across part of the area occurred at least twice during the general deglaciation. The Windermere Interstadial (14-11 ka BP) saw temporarily warmer climate conditions in the study area. Periglacial effects were minor with little or no continuous permafrost, although buried permafrost may have remained from the Late Devensian glacial episode. Extremely severe climatic conditions returned during the Loch Lomond Stadial (ll-10ka BP), but development of glaciers in Cumbria was probably prevented by the shortness of the stadial, except at the highest levels in the uplands as minor corrie glaciers. Permafrost was probably widespread, although the thickness developed may have been curtailed by the brevity of the cold period. An extremely rapid change of climate to temperate conditions warmer than or similar to those of the present day started at 10 ka BP. This is taken to mark the end of the Devensian and the start of the Holocene. The warmest (climatic optimum) conditions were at about 7-5 ka BP.

Quaternary Climate States in the Sellafield Area

At any point in the glacial-interglacial cycle, there is regional variation of climate state governed significantly by latitude, and there is local variation in actual climate conditions within the overall climate state affected by many factors including relief. During the current interglacial, the Sellafield region has a temperate climate state. Because of the strong relief of the Sellafield area, although the coast experiences a temperate climate (average annual temperature 10.2 °C) present day condilions at 800 m above pmsl in the mountains are boreal, i.e. subarctic, with a measured temperature fall with altitude averaging 6.2 °C per 1000m. Similar differences are likely to have existed in the past..

Goodess et al. (1991) showed that over the last 400 ka in the Northern Hemisphere, interglacial conditions with global icesheet extents comparable with those of the modern day existed for only 9% of the time, with a corresponding temperate climate state for the

British Isles including the Sellafield area; at other times, the boreal climate state occurred for 38%, the periglacial climate state for 37%, and the glacial climate state for 16% of this period. The corresponding percentages in the Sellafield area during the last 120ka were: temperate, <27%; glacial 18%; cold (boreal or periglacial) 55%. The higher percentage of the temperate climate state in comparison with the 400 ka interval results in part from the last 120 ka including parts of two interglacials and only one complete glacial.

Climatic conditions during and since the glacial maximum in the UK are reasonably well defined. Prior to the maximum, however, they are less clear. The evidence suggests that for much of the time it was colder than at present, with features characteristic of ground freezing known for a number of sites in deposits of early and mid Devensian age but exact temperatures are difficult to estimate. It seems likely that it may have been rather cold for much of the Devensian; however, the global oxygen isotope record from deep-sea sediments shows that substantial continental ice was not present until the late Devensian. Data from the GRIP ice core from Greenland show that this cold period was not uniformly cold and 24 interstadials are recognised. The sedimentary record from the North Atlantic also shows a number of ice-sheet surge episodes. There is no sedimentary record in the Sellafield area that can be related to these events. Climatic conditions during past interglacial intervals are considered to have been temperate, similar to the present, or slightly warmer at the interglacial maximum.

Data from the analogue sites indicate that the climate during glacial conditions is colder than at the present by 14-22 °C, with much lower precipitation than at present in NW Europe and, given time, the development of a permanent ice sheet. There is no single analogue station that can be closely relied upon to provide data similar to the climate conditions that are expected for the Sellafield area under periglacial conditions. Current periglacial climate types include the high Arctic climate of Spitsbergen, continental climates such as that of central Siberia, alpine climates, and sub-Antarctic island climates such as that of South Georgia, but all of these have different geographical or latitudinal settings from the Sellafield area.

In the sequence of climate states over a glacial-interglacial cycle, the duration of the states can vary. Transitions from one state to another can vary in their abruptness and in the direction to colder or warmer climate states. These characteristics are likely to affect the development of hydrological and hydrogeological conditions. Important hydrogeological parameters may also be very different under a particular climate state, depending on that state's precursor. It appears, for example, that at the close of the last glaciation, the abrupt change to warmer conditions produced a substantial disequilibrium, which may still be affecting hydrogeology at the present day. Hydrogeological conditions during a glaciation that follows a long period of periglacial climate may be strongly influenced by relict permafrost, whereas if a glaciation follows soon after a long period of temperate or boreal climate, little or no permafrost will have built up before the ice-sheet advances over an area.

Quaternary Hydrogeological Changes in the Sellafleld Area

Hydrogeological conditions that would have been affected by the climate changes in the Quaternary include: the base level of the hydrogeological system; the amount of water at the ground surface and the proportion of this that was able to become groundwatcr recharge; the groundwater head and head gradient (distribution of magnitude and direction); the rock permeability; the groundwater chemistry.

These hydrogeological conditions are controlled by relative land and sea level, erosion and deposition, the cold climate processes of permafrost and glaciation, and rock stress changes. The effects of these features and processes on the hydrogeology of the Sellafield area are discussed in the following sections.

Land and Sea Level

For a coastal environment such as that of the Sellafield area, the main control on the base level of the hydrogcological system is the relative sea level. Three important changes have affected land and sea levels and their interrelationships over the Quaternary:

(1) global ocean volumes and corresponding sea level have varied in response to transfer of water to continental ice sheets;

(2) there has been erosion and deposition of material at the Earth's surface;

(3) there has been isostatic adjustment of the Karth's continental crust in response to glacial loading and rock erosion and deposition.

Given these variations, there is a need for a practicable datum surface to which other changes in land surface and sea level can be related. Although the present-day mean sea level can be treated as a theoretical absolute surface relative to the Earth's centre, it is not directly relevant to relative sealevel changes or suitable for use in most numerical modelling of groundwater flow within the rock mass. The most practicable datum is a Rock Mass Datum level, considered to be a fixed surface within the rock mass. The Rock Mass Datum level is taken to be the projection within the rock mass of the surface equivalent to present-day mean sea level. This datum is taken to be fixed within the rock mass and to move with it as the rock mass elevation (measured as from the Earth's centre) varies with time.

Figure : *Sea level sign (2/3 of the way up the cliff face) above Badwater Basin, Death Valley National Park, USA*

In general terms, it is expected that sea levels in the local area will have reflected global (eustatic) levels except during periods when major ice sheets were in the vicinity. In Norway, there is evidence for four periods of major ice-sheet development during the last 115ka. In contrast, there is evidence in Scotland and Ireland, and to a lesser extent in England and Wales, for only two major periods of ice-sheet development during this time interval. The earlier glaciation developed at about 70 ka BP (within the Early Devensian substagc) and the second (Late Devensian) from about 26 ka BP. The effect on the elevation of the rock mass as a result of icesheet loading of the crustal plate, i.e. glacio-isostatic effects, and the extent of an Early Devensian glaciation are unclear. It may be that from about 120 to 26 ka BP, relative sea levels rose and fell in the Irish Sea area in the main according to the global eustatic level. If ice did not approach the Sellafield area, there would have been only limited effects on the elevation of the rock mass in the area. As major ice sheets developed in Scandinavia and northern Britain, in marginal areas there would initially be a minor rise of < 10m, a peripheral forebulge, caused by crustal flexuring and viscous 'flow' of subcrustal material outwards from beneath the ice-sheet load. As the ice sheet approached the area, the crustal rock mass would be increasingly depressed, but with a response time of several thousand years, and so it would not reach the maximum equilibrium depression of about one-third of the ice thickness. For the coastal area at Sellafield, with a maximum ice thickness of about 625 m, the assumed maximum crustal depression would be about 150 m.

***Figure** : The Land Below Sea Level*

When the ice sheet covered the area, from c. 25 to 18 ka BP, the coast probably followed the ice-sheet margin closely. The extent of the sea was therefore defined by the advance and retreat of the ice rather than by relative changes of sea level relative to rock mass. By c. 17-14ka BP, during deglaciation of the northern Irish Sea and the surrounding uplands, relative sea level in Cumbria stood at an unknown, and disputed, height above Rock Mass Datum. Highstands with sealevel elevations of up to 150 m above Rock Mass Datum on both sides of the Irish Sea, proposed by Eyles & McCabe (1989), are not supported by the glacio-hydro-isostatic models of Lambeck or by a recent review of the field evidence. In Cumbria, available sedimentological evidence suggests that the sea may have flooded the present coastal lowlands to about 10-15 m above Rock Mass Datum before about 15 ka BP. Global sea levels at this time were at about -100 m relative to what would have been the position of present mean sea level.

Following the Scottish Readvance, relative sea level fell in the northern Irish Sea basin to reach a lowstand of possibly 55-60 m relative to Rock Mass Datum in the early Holocene, between 10 and 9.5 ka BP, primarily as a result of isostatic recovery of the rock mass exceeding eustatic sealevel rise. Relative sealevel graphs based on onshore data in the Cumbria area do not show such a substantial relative fall, but roughly agree on the timing. Lambeck (1996) suggested that relative sea level remained much higher in the Cumbrian area than further south because of greater isostatic depression of the rock mass, which may reconcile these differences.

During the Holocene, 12 cycles of transgressive and regressive overlap tendencies are recognised in the period from c. 9 ka BP to the present along the coastline of NW England from the Mersey Estuary to the Solway Firth. Transgressive overlaps at various elevations indicate a rapid rise in sea level, known as the Main Flandrian or Holocene Transgression, which concluded at or before about 6 ka BP. In Cumbria, the early Holocene lowstand was followed by this rapid rise in relative sea level. Recently acquired evidence from Nirex studies in the Sellafield area suggests that raised beaches at 6-7 m above OD and raised estuarine flat sediments at 8 m above OD relate to a relative sea-level peak in the mid Holocene (c. 6.5 ka BP) in the NE Irish Sea. No such peak was identified for Maryport (c. 35 km north of Sellafield) or Millom (c. 25 km south), but when present-day tidal ranges of up to 8.2 m are taken into account, a peak in mean sea level of at least 3.5 m above Rock Mass Datum is indicated at c. 6.5 ka BP.

After about 6 ka BP, transgressive-regressive oscillations occurred, not exceeding 4 m. From 5 ka BP to the present day, relative sea level in the Sellafield area has fallen gently to present mean sea level, although small-scale fluctuations of the order of 1 m probably occurred. Similar small-scale fluctuations can confidently be expected to have occurred during the earlier sealevel changes of greater magnitude. Erosion and deposition clayton (1994) considered that the glaciation of northern Britain led to greatly enhanced rates of erosion of the general land surface by ice during the Quaternary. However, the significance of glacial processes as agents of erosion has been disputed by Boardman (1992). He suggests, instead, that a major role of ice during the Quaternary may have been to transport a thick mantle of weathered material that had formed during the Tertiary. Although there is general agreement about the volume of rock moved in producing the present dissected topography of the Cumbrian mountains, opinions differ as to the process and timing.

A simplified description of the geological changes over the last 120 ka in the Sellafield area is necessary for modelling purposes, at least at the present state of model development. A view of the above discussion has been presented by Nirex (1997ft). During the Devensian, the ice sheets may have removed most of the remaining earlier Quaternary sediments and on average eroded a further 20 m of bedrock. This erosion was followed by deposition of an average of 27 m of new superficial deposits (mostly on the lower areas). Nirex (1997ft) suggested that around half of the glacial sediments will be eroded before the arrival of the next ice sheet. As Devensian glacial landforms are still fresh, it is assumed for simplicity that a negligible amount of these has been eroded so far. Net removal of rock from the land surface of 20 m per cycle is equivalent to a 0.1-0.2 mm a"1 averaged over the Quaternary. Similar amounts of erosion and deposition are expected to have occurred during previous glacial-interglacial cycles. Over time, however, these thickness changes are probably largely compensated in terms of ground surface elevation by isostatic changes. Using the Airy isostatic compensation model, and supposing a crustal density of 2670 kg m3 and a mantle density of 3300 kg m', isostatic adjustment compensates for 81% of a given change in crustal thickness. Given the erosion of 20 in of rock, the reduction in ground surface elevation after isostatic adjustment may be expected to be just 3.8 m because the rock mass has risen by 16.2 m.

Permafrost

Permafrost occurs under periglacial conditions and it is likely to have been found at least locally throughout Britain for large parts of any glacial or extreme periglacial episode. In a scoping study, the possible rate of ground freezing at Scllafield was calculated using average rock properties from Nirex (1997c) and the analytical solution of Stefan (1891). The depth of freezing is finally limited by the geothermal gradient through the frozen ground, controlled by a geothermal heat flux estimated to be in the range 60-65 mw m^2. For an annual average ground surface temperature of -5°C (likely to correspond to a lower average annual air temperature), the equilibrium depth of freezing is around 200 m, almost independent of lithology for the thermal properties of rocks at Sellafield given by Nirex (1997r). The calculations show that in the nearsurface environment, the basement rocks (the Borrowdale Volcanic Group) with c. 1% water-filled porosity, freeze much faster than the Triassic sandstones (the Sherwood Sandstone Group) or glacial sediments with c. 20% water-filled porosity, because of the smaller amount of latent heat of crystallisation of water involved. Assuming an effective ground surface temperature of - 5°C and neglecting both geothermal heating and advection, it will take around 2000 years for freezing in the Sherwood Sandstone Group to approach the equilibrium depth controlled by geothermal heating, but the Borrowdale Volcanic Group will freeze to this depth in 200 years. This suggests that an equilibrium depth of permafrost in the Borrowdale Volcanic Group will be achieved in times that are short compared with Quaternary climatic variation, but that equilibrium in the Sherwood Sandstone Group will take a significant fraction of the length of a glacial or periglacial episode.

The review by Watson (1977) of the periglacial environment in Britain concluded that continuous permafrost was present long enough for widespread formation of features such as ice wedges and pingos. Younger (1989) cited evidence for extensive ground freezing in SE England during the Devensian, beyond the limit of the Devensian ice sheet. The present-day boundaries of continuous permafrost in North America and Siberia correspond to average annual air temperatures of colder than -8.5 °C or -7 °C, respectively. This may be compared with average temperatures for Sellafield cold climate analogue stations of O to -2 °C in periglacial conditions, or -3.5 to -12 °C under glacial climate conditions. These climatic ranges Tor the Devensian prior to the last glaciation straddles the conditions for permafrost formation and therefore its possible presence has to be considered.

Where present, permafrost will effectively seal pores and fissures, inhibiting or preventing ground water recharge and discharge. In areas of discontinuous permafrost, localised zones of discharge known as 'taliks' are present where the ground is unfrozen. Taliks may form beneath rivers and lakes, depending on their size. Pitty (1988) considered that even under extreme permafrost development, taliks would be present beneath major rivers in the Sellafield area. Taliks are expected to remain open through the permafrost where the width of a surface water body is approximately twice the permafrost depth.

However, near Fairbanks in Alaska, lakes with a surface area as small as 0.02 km^2 (equivalent diametre 16O m) maintain a through-going talik in thick permafrost. These required dimensions for taliks are large compared with the width of present rivers near Sellafield (the largest, the River Ehen, is around 40 m wide at its mouth), but large ponds such as Braystones Tarn approach this size and larger lakes may have formed in the Sellafield area if rivers were blocked by migrating coastal sand.

There is widespread evidence from a number of sources that the interval around 50 ka BP was warmer than the succeeding and preceding intervals. Outside this period, there is faunal and floral evidence of a cold climate, but the small number of dates from speleothems in northern Hngland suggest that permafrost was not continuous at all times in this period.

Gascoync et al. (1983) suggested an absence of speleothems with dates of 34 13 ka HP, but Gordon et al. (1989) have identified a cluster of dates around 29 ka BP and Atkinson et al. (1986) presented dates as recent as 26 ka BP from locations broadly analogous to Sellafield. It seems likely that it was sufficiently cold to restrict soil carbon dioxide production and ground water circulation at various times prior to the Dimlington Stadial, but that permafrost continuous in space and time was not present until after 26 ka BP.

Glaciation

The hydrogeological signilicancc of continental-scale ice sheets is strongly dependent on the amount of meltwater produced, and how this water migrates away from the ice sheet. Meltwater may be produced both at the ice surface, where solar heating melts potentially large quantities of ice and snow, and by basal and internal melting produced by geothermal heat and from friction as the glacier moves. There are four possible discharge routes for this water:

(1) superficial (subacrial) runoff: this is not thought to play a major part in the drainage of ice sheets;

(2) englacial drainage: the movement of water through crevasses, cavities and tunnels within the ice;

(3) basal drainage: the movement of water between the sole of the ice and the underlying rock, either in thin Alms or via cavities and channels;

(4) recharge to the ground underlying the ice.

The amount of basal and internal meltwater produced depends upon the temperature difference between the ice-sheet surface and its base, and this is controlled by three processes: the conduction of geothermal heat from the bedrock up from the base to the top surface of the glacier; the conduction to the top surface of heat generated by basal and internal friction during movement of the ice sheet; the transfer of heat to new cold ice accumulating on the glacier top surface, from the warmer ice at depth. For the centre of an ice sheet, where lateral movement is negligible at the base of the ice, the temperature difference through the ice is greatest in the absence of new ice arrival, i.e. the base is warmest in these conditions. Arrival of new cold ice cools the system, reducing the temperature difference between the cold atmosphere at the top of the ice and the warmer base of the ice. Arrival of new ice leads to bulk lateral movement away from the centre of the ice sheet. This causes heating within the ice as a result of internal and bed friction, and increases the temperature difference between base and top of the ice. The volumetric rate of ice movement, and therefore usually the velocity and the heating, increases with the distance from the source of the ice sheet. An extensive solution to the movement of an ice sheet was presented by Morland (1994), and a model based on the practical application of these physical principles was described by Boulton & Payne (1992).

Scoping calculations show that for Scottish ice, with a centre over Rannoch Moor in the west Highlands, over 250 km from Sellafield, the temperature difference is of the order of 16 °C for a wide range of ice accumulation rates in excess of 0.2 m a^{-1}. The base of a 700 m thick ice sheet originating from Rannoch Moor will be frozen only if the top surface is colder than - 16 °C (compare this with the range of average temperatures for glacial climates, reduced by 5 °C to allow for 700 m altitude (-8.4 to -17.1 °C); such an ice sheet was probably not frozen to the base. For an ice sheet centred on the Cumbrian Mountains, only 20-30 km distant from Sellafield, the temperature difference is in the range 10-13 °C, and it is more likely that such an ice sheet was frozen at the base for some of the time. The simplified calculation may overestimate the likelihood of basal melting occurring. The boundary dividing Cumbrian Mountains ice from Scottish ice fluctuated around Sellafield. For example, the areas on the coast around the Sellafield works may have been underneath wet-based 'Scottish-Irish Sea ice', and the area inland may have been underneath dry-based 'Cumbrian mountains ice', simultaneously. This would have to be considered for detailed groundwater models of the Sellafield or similar areas.

If the whole of the geothermal heat flux and the whole of the internal heating were available to cause meltwater production, the rate of basal melting at Sellafield can be calculated to be in the range 10-20 mm a^{-1}, potentially available to recharge the rocks underlying the ice sheet. A simple calculation, using a uniform recharge of 20 mm a^{-1} and assuming that the hydraulic properties of the Sherwood Sandstone Group at Sellafield are typical of the Irish Sea Basin (transmissivity 100 m^2 day^{-1} in the top 200 m; this is a conservative assumption, as the rocks of the Basin include mudstone units of assumed low permeability), shows that the hydraulic head at Sellafield to drive this quantity of water through the underlying bedrock to the discharge point at the ice-sheet margin 300 km distant would need to be around 25 km, an obviously unrealistic figure. When the hydraulic head attains a value of c. 0.9 times the ice thickness, the ice will float, opening a water discharge route at the base of the ice and thus dissipating the hydraulic head. For hydraulic heads limited by the likely ice thickness, the maximum recharge is less than 0.4 mm a^{-1}. This implies that alternative discharge routes such as subglacial channels were operative. The approach proposed by Boulton et al. (2001b), of lateral drainage to a subglacial channel network running parallel to the direction of ice flow, results in the whole of the estimated meltwater being able to drain in channels cut through the upper layers of the Sherwood Sandstone, with a channel separation of the order of 100km. Wingfield (1990) has presented evidence of incised channels consistent with the above model. This contrasts with the situation described by van Weert et al. (1997) discussed above.

The possibility of recharge to the ground is also controlled by the presence or absence of permafrost. If permafrost to any considerable depth were present before the advance of an ice sheet, melting of the permafrost by geothermal heating would take hundreds and possibly thousands of years (up to 6700 years for the estimated equilibrium thickness of permafrost in the Sherwood Sandstone Group and using the assumptions discussed above for freezing). Until the permafrost melted fully, there would be only limited possibility of recharge to the ground because most pores and fissures would be closed by ice.

Stress Changes

Differential isostatic depression of the Earth's crust involves bending and generates extensional strains in the upper part of the crust. The constraints of depressing a slab of crust on a spherical Earth have the potential to create transient compressive strains. Scoping calculations have shown that these strains would have been of the order of 10^{-5} at Sellafield for the Devensian ice sheet. Stresses associated with these strains can then be calculated from rock properties.

Loading of the rock mass with significant thicknesses of ice would have produced some vertical stress, although the degree to which this would be effective depends on the ability of pore water to drain in response to the load, resulting in consolidation. The consolidation effect will probably be greater in the vertical than in the horizontal direction as a result of a greater vertical stress increment.

Using available data, Nirex (1999) showed that these effective stresses would have had a negligible effect on the permeability of the Sherwood Sandstone, but would have caused reductions in the permeability of fractures in the Borrowdale Volcanic Group by 50-90%. The calculations are simplified and ignore the hysteresis between stress and fracture aperture, but serve to demonstrate the magnitude of likely effects.

Derived Hydrogeological Conditions for the Sellafield Area During the Late Quaternary

These last have been taken from the range of values presented by Goodess et al. (1991) for the appropriate analogue stations. The 'potential recharge' column represents an estimated value for water available for groundwater recharge. In the absence of ice, this is the remainder from precipitation after satisfying evapourative needs. Not all this water will form groundwater recharge; efficiencies for temperate, boreal and periglacial climates of 80%, 60% and 30%, respectively, are suggested. When permafrost is present, recharge is assumed to be zero; beneath a wet-based ice sheet, available water is assumed to be limited, regionally, to the basal melting rate of about 20 mm a^{-1}.

The main uncertainty in the climatic sequence relates to the latter part of the Mid-Devensian (50-25 ka BP). This was a prolonged period of cold climate, but how cold is not well established. It is therefore necessary to consider two scenarios: one in which the climate from 50 ka to 25 ka BP was similar to that of Siberia at present, with significant permafrost development (the extreme periglacial-glacial analogue stations) consistent with the evidence of permafrost cited by Watson (1977); and one with a climate during this period similar to that of Murmansk at present (one of the periglacial analogues), consistent with the evidence of Atkinson et al. (1986) of ground water circulation resulting in speleothem growth.

To examine this issue from a different viewpoint, climate in the Sellafield area must have passed through an extreme periglacial period, between the temperate climate interstadial prior to 50 ka BP and the maximum of the Late Devensian glaciation. However, if that period were short, then significant permafrost would not have been able to develop in the lowland areas, underlain by the Sherwood Sandstone Group, before icesheet formation. This serves to illustrate that not only the sequence of climate states, but also the order and rapidity of transitions from one to another is important.

The uncertainty in sea level is particularly large between c. 65 and c. 25 ka BP. This increased uncertainty arises from the inconsistency in global sea level between data derived from oxygen isotopes and those from sealevel markers such as submerged speleothems and coral terraces for this period.

The erosion of rock and deposition of new sediment in the Sellafield area, both onshore in the coastal plain and offshore, is also presented. Numerical values are derived from clayton (1994) and Nirex (1997a) Also shown is the resultant change in land surface elevation, taking isostatic adjustment into account (note that no allowance is made for the time lag between erosion and adjustment). It is clear that, although significant erosion has occurred since the Ipswichian, the resultant effect on land elevation is not very marked.

Based on Lambcck (1996), shows a relative sea level at +20 m for the Sellafield area at 22 ka BF during the main Late Devensian Glaciation, which requires around 150m of isostatic depression. The estimate by Lambeck (1996) of crustal depression considers the effects of an elastic crust over a viscous mantle and does not reach equilibrium isostatic depression in the Sellafield area: the depression of 15Om is caused by around 500-700 m of ice in the area, with the ice surface at about 625 m relative to the Rock Mass Datum at the location of the present Sellafield works. This would leave Scafell Pikes and other peaks protruding as nunataks, consistent with the observations of Lamb & Ballantyne (1998).

Implications for the Hydrogeological System

The processes discussed above affect both the hydraulic head of groundwater and its gradient, which, together with permeability, determines the flow of groundwater. Hydraulic head gradient and flow are interdependent. The presence or absence of ice may have a significant effect on the availability of water and thus on hydraulic head gradient. Gradient may be controlled externally by a glacially imposed distribution of hydraulic head. Absolute values of hydraulic head are affected both by the integrated effect of the gradient and by imposed boundary values.

Regional discharge levels, determining the downstream boundary hydraulic head, will generally have varied with relative sea level in the absence of glaciation, although during periods of extremely low sea level, when the coastline would have been far removed from Sellafield, the aquifer drainage point (lowest outcrop or subcrop of the sandstone not confined by the overlying mudstones) may have been above sea level.

When ice sheets were absent from the area, horizontal hydraulic gradients would have similar magnitudes to those at the present day, although during periods of low precipitation and consequently restricted recharge it is likely that they would decrease, at least in the lowlands. Under wetter conditions, hydraulic gradients could increase until the water table reaches the ground surface, when the gradient becomes limited by surface slope. This condition already obtains in the upland regions near Sellafield under the present climate regime. The present-day hydraulic gradient is around 0.02, and it would have decreased in colder and drier periods to 0.01 or less. Given the evidence for rapid climate change resulting in the Dansgaard-Oeschger cycles and associated Heinrich events, it seems likely that hydraulic gradients could have fluctuated over this range many times over the long period preceding the main late Devensian glaciation.

Groundwater gradients may take extremely long periods to equilibrate to significant changes in surface pressures, depending on the hydraulic diffusivity of the rock mass (the ratio of hydraulic conductivity to storage coefficient). Hydraulic diffusivities and groundwater heads were measured over 50 m lengths down the full length of the deep boreholes as described by Sutton (1996) and have been summarised by Nirex (1998). Long equilibration times are particularly likely in the low-permeability Borrowdale Volcanic Group at depth where, for example, the anomalous present-day vertical head gradient of 0.05 to a depth of over 1540m below present sea level in Borehole 2 may be due to continuing disequilibrium with modern surface conditions. Simple calculations using an analytical solution of the ID diffusion equation suggest that heads in the Borrowdale Volcanic Group at depth might not respond fully (90%) to changes in sea level or recharge until 100 ka later. Similar calculations show that the permeable near-surface sandstones of the Sherwood Sandstone Group would react rather quickly (within years) to such changes.

Hydraulic gradient directions in the Sellah'cld area may have changed as a result of the changing configuration of ice sheets towards the end of the glacial period. If a lobe of Scottish ice persisted in the Irish Sea over the western part of the area, while the Cumbrian mountains ice had retreated to the mountains, locally hydraulic gradients might have been from west to east, away from the remaining ice. If Irish Sea-Scottish ice covered the whole coastal plain and was confluent with the Cumbrian mountains ice sheet, then hydraulic gradients could have been approximately from north to south, controlled by the ice-sheet surface, or from east to west, controlled by drainage to northsouth subglacial channels. Present-day ground water hydraulic gradients are to the WSW.

A considerable amount of water issues from active ice-sheet margins, as here ice-sheet advance is controlled by melting. Boulton & Caban (1995) have documented evidence of failure of unconsolidated sediments in such situations, and this probably places an upper limit on the hydraulic gradient that can be achieved of around two, at the ice-sheet boundary. These represent the cases corresponding to development or non-development of significant permafrost prior to the Dimlington Stadial. These alternatives are discussed in the following sections.

Case A: Permafrost not Significantly Developed Prior to the Dimlington Stadial

The balance of the evidence, especially the study by Gordon et al. (1989), appears to suggest that continuous permafrost was absent during the Devensian. Groundwater heads would be relatively low because of the low available recharge, from low precipitation and partially frozen ground. Groundwater heads would be expected to decrease towards the onset of ice-sheet development because of the formation of more extensive permafrost that would have further reduced recharge.

The review of ice-sheet processes earlier recognised the uncertainty in whether the main Late Devensian ice sheet would have been wet-based or dry-based in the Sellafield area. Once wet-based ice covered the area the limited permafrost would have melted as a result of geothermal heating. Within a relatively short period, groundwater heads would rise locally towards a value similar to that required to float the ice (because at that point it is considered that melt water would find alternative discharge routes at the ice-sheet sole) but would be spatially variable in the presence of a subglacial drainage channel system. Under wet-based conditions, the vertical component of the hydraulic head gradient will be downwards, at least initially. Effective stresses in the rock mass would remain low in areas of high fluid pressure. If dry-based ice covered the area, the aquifer would have become inactive (case A2). Under these conditions, it is more likely that excess pressure produced by loading will dissipate, especially from the sandstone. Effective stresses in the rock may therefore be high. Later, as the permafrost melted and the ice sheet warmed as a result of climatic amelioration, the ice would have become wetbased. It is assumed that a dynamic equilibrium was achieved where the ice was not quite floating; that is, hydraulic head at base of ice would be similar to ice load, as the basal melting of about 20 mm a^{-1} could not be transmitted as long-distance groundwater flow through the rocks in the Sellafield area as discussed above.

The horizontal hydraulic gradient would have followed the upper unbroken line, decreasing gradually from a value similar to that of today as recharge became more limited, then rising briefly when the ice front passed over, before dropping to a value controlled by water pressures within the ice sheet and hence by the gradient of the ice-sheet surface, which is expected to have been of the order of 0.001. Locally higher gradients may have formed near subglacial drainage channels.

Case B: Permafrost Well Developed Prior to the Dimlington Stadial

The ice would have advanced over ground frozen to a thickness of up to 200 m. With this thickness of permafrost, taliks completely penetrating the permafrost are likely to have been rare in the vicinity of Sellafield. Groundwater hydraulic heads would have been very low because of the lack of recharge. Simple calculations show that the Sherwood Sandstone aquifer would have taken less than 1000 years to drain after the cessation of recharge.

If the ice was wet-based, the permafrost would have prevented meltwater and therefore hydraulic head from being transmitted to the groundwater in the rocks beneath the ice for much of the glacial period, as the rate of permafrost decay beneath the ice was too slow. If the ice sheet remained frozen throughout, it would not allow development of a wet base, and the upper layer of the aquifer, too, might remain frozen. Therefore, the results of a wet base or a dry base are the same if the upper part of the aquifer is frozen: it is not possible to transmit water from the glacier into the rock mass. However, a load equal to the weight of the glacier is imposed on the rock mass and this will tend to increase the hydraulic head by compressing the pore space. Whether this translates into an increase in effective stress in the rock mass depends on whether groundwater in the pores can drain up to the ice-sheet base: such drainage is likely to have been impeded by permafrost. Initially, at least, groundwater head would increase. Head gradients are likely to have had a large upward component in such a situation.

Eventually, the ice would become wet based, the permafrost would melt and the hydraulic regime would converge with that of the 'no permafrost' case (B2).

The prevention of recharge during permafrost conditions would have resulted in a gradual drainage of the aquifer and reduction of the horizontal hydraulic gradient. On the arrival of the ice, the horizontal component would have been controlled by changes in loading associated with large-scale variations in ice thickness, and would be fairly small, despite a probable increase in pressure, and the aquifer would have been very largely inactive. After basal melting, gradients would then remain very low. During the transition from low head-low gradient as a result of permafrost, to high head-low gradient controlled by water within the glacier, transient large hydraulic gradients may have occurred as a result of irregular melting.

Groundwater hydraulic head would have remained high, dependent principally on the thickness of ice, until the retreat of the ice sheet.

Ice-sheet Retreat

As the ice decayed, basal drainage probably became well integrated, at least during summer, and hydraulic heads in the ice therefore would fall considerably. Hydraulic heads may have risen each winter as pores and cracks in the ice refroze and drainage became less efficient. This would have given rise to large seasonal head variations, rather more rapid than can be shown. Permafrost would have decayed by this time, so these hydraulic head changes would have been communicated to the groundwater in underlying rocks. During the two periods of readvance, new ice build-ups would have reintroduced higher hydraulic heads from the water table in the ice sheet, until the final melting of the ice caps and onset of the Windermere Interstadial. Recharge and groundwater head since that time are thought to have been similar to those at the present, although somewhat reduced in the cold and presumed dry period of the Loch Lomond Stadial.

During retreat, potentially very large hydraulic gradients could have been applied close to the ice margin and huge quantities of meltwater escaped (perhaps > $3000 m^3 s^{-1}$, for short periods; Piotrowski 1997). Gradients may have been limited only by the strength of the rock mass. Periodic oscillations of the ice margin during ice-sheet decay, and the development of jökulhlaup (Icelandic for 'glacial outburst resulting in flooding') or similar phenomena would give rise to great variation in hydraulic gradient over this time.

Evidence for Hydrogeological Conditions during the Quaternary

Groundwater Chemistry

The changing pattern of hydraulic heads and gradients leaves little evidence for geologists. However, the chemistry of the groundwater may be distinctive and this persists for examination. In the sequence of events described above for the last interglacial-glacial cycle, the Sellafield area, except perhaps the extreme coastal margins, was always subject to some form of meteoric regime and therefore any new water that recharged the groundwater system would have been fresh. Glacier dynamics are such that basal meltwater is derived from precipitation that originally fell near the source of the glacier, at relatively high altitude. From the combination of cold climate and high altitude, it would be expected that this meltwater would have a high dissolved oxygen content and an unusually 'light' stable hydrogen and oxygen isotope composition, which is confirmed by observations of a modern high-latitude glacier. Permafrost processes can produce changes in groundwater chemistry; as groundwater freezes, solute concentrations (low in the ice) increase in the remaining groundwater and may form a saline zone at the margins or base of the frozen area.

The presence of a saline transition zone separating fresh water above from saline water below in the Sellafield area has been described by Bath el al. (1996). The position of such a transition zone is determined by hydraulic potentials and its position would be expected to be sensitive to changes in the groundwater flow regime.

The model of groundwater system evolution presented above suggests that there should be ground waters with a distinctively 'light' stable isotope composition, water of marine origin should be absent, and that the saline transition zone will have moved during the glacial-interglacial cycle.

The groundwater sampling programme undertaken during the Nirex investigation of the Sellafield area was extensive, both spatially and with depth. Isotopically 'light' fresh water was identified at several hundred metres depth in the sandstone aquifer, which may indicate older freshwater recharged at lower temperatures than those of the present. However, extremely 'light' water, which might result from the melting of high-altitude snow transported at the base of a glacier, has not been observed at any point in the Sellafield groundwater system. No marine saline water was identified at depth.

Pore and Fracture Fills

Changes in the chemistry of groundwater have left a record in the mineralogy of pore and fracture-filling minerals. A sequence of mineral infilling episodes was recognised and the youngest two, ME8 and ME9, are of particular interest for Quaternary studies.

ME8 is the result of oxidative processes, requiring active circulation of groundwater of meteoric origin. These processes can be observed occurring in the near-surface at present and have also affected late Devensian and prelate Devensian glacial deposits.

ME9 represents mineralisation occurring in the deeper reducing ground waters. It is dominated by the deposition of calcite with minor pyrite, marcasite, barite, an hydrite and gypsum. There is evidence of it overlapping in time with ME8. The variation in the depth of penetration of ME8 mineralisation and alteration in fractures shows a relationship between the saline transition zone and the maximum extent of ME8 mineralisation: the base of the zone occupied by ME8 mineralisation follows approximately the line of the saline transition zone. ME8 oxyhydroxides were observed to be included within the cores of ME9 calcite, just above the saline transition zone. As ME8 minerals form principally by the dissolution of pre-existing calcite and dolomite, the overprinting of ME8 mineralisation by ME9 calcite within this interval implies that the maximum depth of penetration of ME8 reflects an earlier hydrogeological regime, rather than the present one.

The observations suggest that in the late Tertiary, or early in the Quaternary, oxidising and carbonate-leaching ground waters penetrated to a greater depth than at present (probably limited by the permeability contrasts in the lowest beds of the sandstone aquifer unit). However, the groundwater regime then responded to changes during the Quaternary and the most recent mineralogical phenomena suggest that the 'redox front' and the calcite dissolution-precipitation front have risen in elevation during later stages of the Quaternary. These conclusions are also consistent with the direction of movement of the saline transition zone inferred from observations of changes in ME9 calcite morphology.

^{230}Th/U isotopic dating of whole crystals gave ages from 17 ka BP to > 300 ka BP on samples of ME9 calcite, indicating that some of it is Pleistocene and possibly Devensian in age. The fact that these are whole-crystal ages does not exclude that the rims of the crystals are much younger than the cores and could even be modern. There is much evidence for zoning in these calcite crystals and also some evidence for breaks in deposition and even temporary dissolution. This indicates time variation in the controlling processes, which could be the result of Quaternary climate change, although there is no direct evidence for this.

Systematic changes with depth in the morphology of ME9 calcite have been observed in all the Sellafield boreholes that have been studied and can be correlated with groundwater salinity. At shallow levels, c-axis flattened 'nailhead' crystals are associated with freshwater, whereas deeper calcite crystals typically exhibit c-axis elongation ('dog-tooth' crystals) in contact with saline water. Between them, study of overgrowth history in crystals reveals a history of changing morphology deduced to reflect changing salinity. Thus it is possible to deduce the range of elevation of the saline transition zone over a significant period of the Quaternary, in comparison with its present elevation. The detailed sequence of morphological changes of calcite crystals possibly provides a record of the elevation of the saline transition zone, with which hydrogeological models must be consistent. The broad correlation between the mineralogical changes and present-day changes in groundwater chemistry suggests that the groundwater system has been robust to external changes caused by interglacial-glacial processes during the late Quaternary. This contrasts with the early thinking of Heathcote (1997) and Boulton et al. (1995) discussed in the introduction.

Climate changes over the last 120 ka have the potential to affect considerably the groundwater regime at a coastal location such as Sellafield. Different aspects of the changes may affect groundwater flux (in magnitude and direction) and groundwater head. Except at great depth, the evidence of these changes having occurred will be indirect (for instance, hydrochemical and mineralogical). Although present understanding of subglacial drainage implies that large amounts of fresh glacial meltwater could move through a highly permeable aquifer under glacial maximum conditions, there is little evidence of this having occurred at Sellafield: distinctive isotopically light groundwater derived from subglacial melting is not present, and there is evidence that the transition between shallow freshwater and deep saline water has not moved greatly during the last 100 ka or more. Any changes may have been confined to the upper layers of the groundwater system, and have been removed by the present-day active flow regime.

An attempt has been made to establish a consistent set of boundary conditions, which could be used to control a detailed numerical hydrogeological model, to investigate the effect of late Quaternary climate changes on the groundwater system. These boundary conditions are complex and vary in time and space. Hydrogeological properties of the deeper fractured rocks may have varied significantly during glaciation. Lastly, there is a general upward movement of the rock mass in response to erosion.

8

Hydropolitics and Equity Measures for Water-sharing Agreements

Because of water's preeminent role in survival from an individual's biology to a nation's economy political conflicts over international water resources tend to be particularly contentious. The intensity of a water conflict can be exacerbated by a number of factors, including a region's geographic, geopolitic, or hydropolitic landscape. Water conflicts are especially bitter, for example, where the climate is arid, where the co-riparians of regional waterways are otherwise engaged in political confrontation, or where the population's water demand is already approaching or surpassing annual supply.

The watersheds of the arid and volatile Middle East provide settings for water conflicts of extreme intensity, in that each of the three major waterways the Nile, the Jordan, and the Tigris-Euphrates systems have elements of all of these exacerbating factors. In fact, as will be discussed below, scarce water resources have already been at the heart of much of the bitter, occasionally armed, conflict endemic to the region. It is of little wonder that Boutros Boutros-Ghali, currently Secretary-General of the United Nations, in the past suggested that a future war in the Middle East may be fought over water.

One can find room for optimism, however, in the fact that the same characteristics of water resources which fuel conflict can, if managed carefully, induce cooperation in an environment of hostility. According to Frey and Naff, "precisely because it is essential to life and so highly charged, water can perhaps even tends to produce cooperation even in the absence of trust between concerned actors." At the heart of water conflict management, is the question 'equity.'

A vague and relative term in any event, criteria for equity are particularly difficult to determine in water conflicts, where international legal guidelines are poorly developed. However, application of an 'equitable' water-sharing agreement along the volatile waterways of the Middle East is a prerequisite to hydropolitical stability which, finally, could help propel political forces away from conflict in favour of cooperation.

This chapter explores the question of equity measures for water-sharing agreements in the context of Middle East hydropolitics and is divided into three parts. The next section provides a brief summary of the hydropolitical history of the major waterways of the Middle East, focusing on international instances of hydro-conflict or, alternately and more rarely, hydro-cooperation. Following is a section which offers various measures of water-sharing equity which have been proposed in different settings and academic disciplines, and which may be useful to the process of water conflict resolution.

Water Conflict and Cooperation ill the Middle East

Living as they do in a transition zone between Mediterranean, subtropical and arid climates, the people in and around the major watersheds of the Middle East have always been aware of the limits imposed by scarce water resources. Settlements sprang up in fertile valleys or near large, permanent wells, and trade routes were established from oasis to oasis. In ancient times, cycles of weather patterns had occasionally profound effects on the course of history. For example, recent research suggests that climatic changes 10,000 years ago, which caused the average weather patterns around the Dead Sea to become warmer and drier, may have been an important factor in the birth of agriculture in the region.

The fluctuating waters of the ancient Middle East have given rise to legend, extensive water law, and the roots of modern hydrology: The flood experienced by Noah is thought to have centred its devastation around the Babylonian city of Ur, submerging the southern part of the Euphrates for about 150 days, while the code of King Hammurabi contains as many as 300 sections dealing with irrigation. The practice of field surveying was invented to help harness the flooding Nile. In addition, the waters of the region were occasionally intertwined with military strategy as, for instance, when Joshua directed his priests to stem the Jordan's flow with the power of the Ark of the Covenant while he and his army marched across the dry riverbed to attack Jericho.

In the centuries since, the inhabitants of the region and the conquering nations which came and went have lived mostly within the limits of their water resources, using combinations of surface water and well water for survival and livelihood. It is in the beginning of this century, as the competing nationalisms of Jews and Arabs began to re-emerge on the ruins of the Ottoman Empire, that the quest for resources took on a new and vital dimension. In the years that followed World War I, the location of water resources influenced the boundaries first between the British and French mandate powers which acquired control over the region, then between the states which developed subsequently. The Zionist border formulation for a 'national home' presented at the Paris Peace Talks in 1919, for example, was determined by three criteria: historic, strategic, and economic.

Economic considerations were defined almost entirely by water resources. The entire Zionist program of immigration and settlement required water for large-scale irrigation and, in a land with no fossil fuels, for hydro-power. The development plans, and the boundaries which were required, were, "completely dependent" on the acquisition of the "headwaters of the Jordan, the Litani River, the snows of Hermon, the Yarmuk and its tributaries, and the Jabbok."

National interests along the Nile likewise were being informed by water resources development at the same time. In the early 1900's, a relative shortage of cotton on the world market put pressure on Egypt and the Sudan, then under a British-Egyptian condominium, to turn to this summer crop, requiring perennial irrigation over the traditional flood-fed methods. The need for summer water and flood control drove an intensive period of water development along the Nile, with proponents of Egyptian and Sudanese interests occasionally clashing within the British foreign office over whether the emphasis for development ought to be further upstream or down.

The most extensive scheme for comprehensive water development along the Nile, proposed in 1920 and now known as the Century Storage Scheme, included a storage facility on the Uganda-Sudan border, a dam at Sennar to irrigate the Gezira region south of Khartoum, and another dam on the White Nile to hold summer flood water for Egypt. The plan worried some Egyptians, and was criticised by nationalists, because all the major control structures would have been beyond Egyptian territory and authority. A Nile Project Commission mediated the immediate conflict in 1920 although no final agreement was reached at the time. In 1925, however, a new commission made recommendations based on the 1920 estimates which would lead finally to the Nile Waters Agreement of 1929. Four thousand MCM/yr. was allocated to Sudan but the entire timely flow (from January 20 to July 15) and a total annual amount of 48,000 MCM/ yr. was reserved for Egypt.

Between the World Wars, water became the focus of the greater political argument over how to develop the budding states around the Jordan watershed, particularly Israel and Jordan, and what the 'economic absorptive capacity' would be for immigration. Development plans included the Ionides (1939) plan, a British study which suggested that water would be a limiting factor for any additional immigration to Palestine, and the Lowdermilk (1944) plan which suggested in contrast that, with proper water management, resources would be generated for four million refugees in addition to the 1.8 million Arabs and Jews living in Palestine at the time. British policymakers came down on the side of the Ionides Plan, invoking 'economic absorptive capacity' to limit Jewish immigration and land transfers for the duration of World War II.

As the borders of the new states were defined, sometimes by warfare, in the 1950s and 1960s, each country began to develop its own water resources unilaterally. On the Jordan River, the legacy of the Mandates, and the 1948 war, was a river divided in a manner in which conflict over water resource development was inevitable. By the early 1950s, Arab states were discussing organised exploitation of two northern sources of the Jordan the Hasbani and the Banias. The Israelis also made public their All Israel Plan, which included the draining of Huleh Lake and swamps, diversion of the northern Jordan River and construction of a carrier to the coastal plain and Negev Desert the first out-of-basin transfer for the watershed.

Jordan, in 1951, announced a plan to irrigate the East Ghor of the Jordan Valley by tapping the Yarmuk. At Jordan's announcement, Israel closed the gates of an existing dam south of the Sea of Galilee and began draining the Huleh swamps, which lay within the demilitarised zone with Syria. These actions led to a series of border skirmishes between Israel and Syria which escalated over the summer of 1951. In July, 1953, Israel began construction on the intake of its National Water Carrier at the Daughters of Jacob Bridge north of the Sea of Galilee and in the demilitarised zone. Syria deployed its armed forces along the border and artillery units opened fire on the construction and engineering sites. Syria also protested to the U.N.

and, though a 1954 resolution for the resumption of work by Israel carried majority, the USSR vetoed the resolution. The Israelis then moved the intake to its current site at Eshed Kinrot on the northwestern shore of the Sea of Galilee.

Against this tense background, President Dwight Eisenhower sent his special envoy Eric Johnston to the Middle East in October 1953 to try to mediate a comprehensive settlement of the Jordan River system allocations. Johnston's initial proposals were based on a study carried out by Charles Main and the Tennessee Valley Authority (TVA) at the request of the U.N. to develop the area's water resources and to provide for refugee resettlement.

The major features of the Main Plan included small dams on the Hasbani, Dan and Banias, a medium size (175 MCM storage) dam at Maqarin, additional storage in the Sea of Galilee, and gravity-flow canals down both sides of the Jordan Valley. The Main Plan excluded the Litani River and described only in-basin use of the Jordan River water, although it concedes that, "it is recognised that each of these countries may have different ideas about the specific areas within their boundaries to which these waters might be directed." Preliminary allocations gave Israel 394 MCM/yr., Jordan 774 MCM/yr., and Syria 45 MCM/yr.

Both Israel and a united Arab League Technical Committee responded with their own counterproposals, and Johnston worked until the end of 1955 to reconcile these proposals in a Unified Plan amenable to all of the states involved. In the Unified Plan, Johnston accomplished no small degree of compromise. Though they had not met face to face for these negotiations, all states agreed on the need for a regional approach. Israel gave up on integration of the Litani and the Arabs agreed to allow out-of-basin transfer. The Arabs objected, but finally agreed, to storage at both the Maqarin Dam and the Sea of Galilee so long as neither side would have physical control over the share available to the other. Israel objected, but finally agreed, to international supervision of withdrawals and construction. Allocations under the Unified Plan, later known as the Johnston Plan, included 400 MCM/yr. to Israel, 720 MCM/yr. to Jordan, 132 MCM/yr. to Syria and 35 MCM/yr. to Lebanon.

The technical committees from both sides accepted the Unified Plan but forward momentum died out in the political realm, and the Plan was never ratified. Nevertheless, Israel and Jordan have generally adhered to the Johnston allocations and technical representatives from both countries have met from that time until the present two or three times a year at "Picnic Table Talks," named for the site at the confluence of the Yarmuk and Jordan Rivers where the meetings are held, to discuss flow rates and allocations.

To the west at the same time, hydrologic developments along the Nile were caught up in the nationalist movements of both Egypt and Sudan. The Aswan High Dam, with a projected storage capacity of 156,000 MCM/yr., was proposed in 1952 by the new Egyptian government, but debate over whether it was to be built as a unilateral Egyptian project or as a cooperative project with Sudan kept Sudan out of negotiations until 1954. The negotiations which ensued, and carried out with Sudan's struggle for independence as a back-drop, focused not only on what each country's legitimate allocation would be, but whether the dam was even the most efficient method of harnessing the waters of the Nile.

Sudan, hoping to guarantee its own future water needs, advocated the planning envisioned in the Century Storage Scheme, with its emphasis on upstream controls, and balked at Egyptian desires for water additional to its 1929 allocations. Egypt based these new allocations of 62,000 MCM/yr. of the river's projected net discharge of 70,000 MCM/yr. on the basis of the "primary needs" of its larger population and on the lack of any other water supplies. Negotiations were broken off and relations threatened to degrade into military confrontation in 1958 when Egypt sent an unsuccessful expedition into territory in dispute between the two countries.

Sudan attained independence in 1956, but it was with the military regime which gained power in 1958 that Egypt adopted more conciliatory tone in the negotiations which resumed, and with which the Nile Water Treaty was signed on November 8, 1959. The total recognised water rights for each state of 55,500 MCM/yr. for Egypt and 18,500 MCM/yr. for Sudan are still upheld today. The treaty also provided for a Sudanese water "loan" to Egypt of up to 1,500 MCM/yr. through 1977; for a Permanent Joint Technical Committee to resolve disputes and jointly review claims by any other riparian; and for equal sharing of future increases in the yield of the Nile. Egypt and Sudan agreed that the combined needs of other riparians, at the time still under British rule, would not exceed 1,000-2,000 MCM/yr. Ethiopia, which until then had not been a major player in Nile hydropolitics, served notice in 1957 that it would pursue unilateral development of the Nile water resources within its territory, estimated at 75-85% of the annual flow, and suggestions were made recently that Ethiopia may eventually claim up to 40,000 MCM/yr. for its irrigation needs both within and outside of the Nile watershed. No other state riparian to the Nile has ever exercised a legal claim to the waters allocated in the 1959 treaty.

As each state developed its water resources unilaterally, their plans began to overlap. By 1964, for instance, Israel had completed enough of its National Water Carrier that actual diversions from the Jordan River basin to the coastal plain and the Negev were imminent. Although Jordan was also about to begin extracting Yarmuk water for its East Ghor Canal, it was the Israeli diversion which prompted President Nasser to call for the First Arab Summit in January, 1964, including heads of state from the region and North Africa, specifically to discuss a joint strategy on water.

The options presented at the Summit were to complain to the U.N., divert the upper Jordan tributaries into Arab states, as had been discussed by Syria and Jordan since 1953, or to go to war. The decision to divert the rivers prevailed at a Second Summit in September, 1964, and the Arab states agreed to finance a Headwater Diversion project in Lebanon and Syria and to help Jordan build a dam on the Yarmuk. They also made tentative military plans to defend the diversion project.

In 1964, Israel began withdrawing 320 MCM/yr. of Jordan water for its National Water Carrier and Jordan completed a major phase of its East Ghor Canal. In 1965, the Arab states began construction of their Headwater Diversion Plan to prevent the Jordan headwaters from reaching Israel. The plan was to divert the Hasbani into the Litani in Lebanon and the Banias into the Yarmuk where it would be impounded for Jordan and Syria by a dam at Mukheiba. The diversion would divert up to 125 MCM/yr., cut by 35% the installed capacity of the Israeli Carrier, and increase the salinity in the Sea of Galilee by 60 ppm. In March, May, and August of 1965, the Israeli army attacked the diversion works in Syria. These events set off what has been called "a prolonged chain reaction of border violence that linked directly to the events that led to the (June 1967) war. Border incidents continued between Israel and Syria, finally triggering air battles in July 1966 and April 1967, and finally to all-out war in June 1967.

In the territorial gains and improvements in geostrategic positioning which Israel achieved in the war, Israel also improved its "hydrostrategic" position. This was largely due to its acquisition of the Golan Heights. With the Golan Heights, it now held all of the headwaters of the Jordan, with the exception of a section of the Hasbani, and an overlook over much of the Yarmuk, together making the Headwaters Diversion impossible. The West Bank not only provided riparian access to the entire length of the Jordan River, but it overlay three major aquifers, two of which Israel had been tapping into from its side of the Green Line since 1955. Jordan had planned to transport 70-150 MCM/yr. from the Yarmuk River to the West Bank. These plans, too, were abandoned.

In the years since, increased integration of the West Bank and Gaza into the economic and hydrologic networks of Israel has led to increasing hydropolitical tensions. As mentioned above, when Israel took control of the West Bank and Gaza in 1967, the territory captured included the recharge areas for aquifers which follow west and northwest from the West Bank into Israel, and east to the Jordan Valley. The entire renewable recharge of these first two aquifers is already being exploited and the recharge of the third is close to being depleted as well.

In the years of Israeli occupation, a growing West Bank and Gaza population, along with burgeoning Jewish settlements, has increased the burden on the limited groundwater supply, resulting in an exacerbation of already tense political relations. Palestinians have objected strenuously to Israel control of local water resources and to settlement development, which they see as being at their territorial and hydrologic expense. Israeli authorities view hydrologic control in the West Bank as defensive. With about 30% of Israeli water originating on the West Bank, the Israelis perceive the necessity to limit groundwater exploitation in these territories in order to protect the resources themselves, and their wells from saltwater intrusion.

Hydropolitical tensions have not been limited to the Nile and the Jordan basins. In 1975, unilateral water developments came very close to leading to warfare along the Euphrates. The three riparians to the river Turkey, Syria and Iraq had been coexisting with varying degrees of hydropolitical tension through the 1960's. At that time, population pressures drove unilateral developments, particularly in southern Anatolia, with the Keban Dam (1965-73), and in Syria, with the Tabqa Dam (1968-73).

Bilateral and tripartite meetings, occasionally with Soviet involvement, had been carried out between three riparians since the mid 1960's, although no formal agreements had been reached by the time the Keban and Tabqa dams began to fill late in 1973, resulting in decreased flow down-stream. In mid 1974, Syria agreed to an Iraqi request that Syria allow an additional flow of 200 MCM/yr. from Tabqa. The following year, however, the Iraqis claimed that the flow had been dropped form the normal 920 m sup3/sec to an "intolerable" 197 m sup3 /sec, and asked that the Arab League intervene. The Syrians claimed that less than half the river's normal flow had reached its borders that year and, after a barrage of mutually hostile statements, pulled out of an Arab League technical committee formed to mediate the conflict. In May 1975, Syria closed its airspace to Iraqi flights and both Syria and Iraq reportedly transferred troops to their mutual border. Only mediation on the part of Saudi Arabia was able to break the increasing tension, and on June 3, the parties arrived at an agreement which averted the impending violence. Although the terms of the agreement were not made public, Naff and Matson cite Iraqi sources as privately stating that the agreement called for Syria to keep 40% of the flow of the Euphrates within it borders, and to allow the remaining 60% through to Iraq.

By 1991, several events combined to shift the emphasis on the potential for 'hydro-conflict' in the Middle East to the potential for 'hydro-cooperation.' The first event was natural, but limited to the Jordan basin. Three years of below-average rainfall caused a dramatic tightening in the water management practices of each of the riparians, including rationing, cut-backs to agriculture by as much as 30%, and restructuring of water pricing and allocations. Although these steps placed short-term hardships on those affected, they also showed that, for years of normal rainfall, there was still some flexibility in the system. Most water decision-makers agree that these steps, particularly regarding pricing practices and allocations to agriculture, were long overdue.

The next series of events were geopolitical, and region-wide, in nature. The Gulf War in 1990 and the collapse of the Soviet Union caused a realignment of political alliances in the Mideast which finally made possible the first public face-to-face peace talks between Arabs and Israelis, in Madrid on October 30, 1991. During the bilateral negotiations between Israel and each of its neighbours, it was agreed that a second track be established for multilateral negotiations on five subjects deemed 'regional,' including water resources. Although the pace of the peace talks has been at times arduously slow, a venue does finally exist where grievances can be aired and the issue of water-sharing equity can be tackled. This, in itself, may help prevent the some of the pressures that have historically led to some of the most bitter water conflicts in the world.

Measures of Water-sharing Equity

One problem at the heart of Middle East water conflicts is the fact that there is no internationally accepted definition of water-sharing equity. International water law is ambiguous and often contradictory, and no mechanism exists to enforce principles which are agreed-upon. This section describes some measures of water-sharing equity which do exist, their strengths, and their weaknesses.

International Water Law

According to Cano, international water law did not substantially begin to formulate until after World War I. Since that time, organs of international law have tried to provide a framework for increasingly intensive water use. The concept of a 'drainage basin,' for example, was accepted by the International Law Association in the Helsinki Rules of 1966, which also provides guidelines for "reasonable and equitable" sharing of a common waterway. Article IV of the Helsinki Rules describes the overriding principle:

Each Basin State is entitled, within its territory, to reasonable and equitable share in the beneficial uses of the waters of an international drainage basin. Article V lists no fewer than eleven factors which must be taken into account in defining what is "reasonable and equitable." There is no hierarchy to these components of "reasonable use;" rather they are to be considered as a whole. One important shift in legal thinking in the Helsinki Rules is that they address right to "beneficial use" of water, rather that to water per se.

The International Law Commission, a body of the United Nations, was directed by the General Assembly in 1970 to study "Codification of the Law on Water Courses for Purposes other than Navigation." The general principles being codified include:

1) Common water resources are to be shared equitably between the states entitled to use them, with related corollaries of (a) limited sovereignty, (b) duty to cooperate in development, and (c) protection of common resources.
2) States are responsible for substantial transboundary injury originating in their respective territories.

The problems arise when attempts are made to apply this reasonable but vague language to specific water conflicts. In the Middle East, for example, riparian positions and consequent legal rights shift with changing borders, many of which are still not recognised by the world community. Furthermore, international law only concerns itself with the rights and responsibilities of stares. Some political entities who might claim water rights, therefore, would not be represented, such as the Palestinians along the Jordan or the Kurds along the Euphrates.

International law seeks to develop general principles which can then be applied to specific problems. It is testimony to the difficulty of marrying legal and hydrologic intricacies that the International Law Commission, despite an additional call for codification at the U.N. Water Conference at Mar de Plata in 1977, has not yet completed its task. After twenty years and nine reports, only several articles have been provisionally approved. Even once the details are worked through, the principles would not have the force of law until approved by the U.N. General Assembly. Even then, cases are heard by the International Court of Justice only with the consent of the parties involved, and no practical enforcement mechanism exists to back up the Court's findings, except in the most extreme cases. A state with pressing national interests can therefore disclaim entirely the court's jurisdiction or findings.

Needs-Based Equity

Many of the common claims for water rights are based either on geography, i.e. from where a river or aquifer originates and how much of that territory falls within a certain state, or on chronology, i.e. who has been using the water the longest. The extreme positions of either definition have been referred to as 'the doctrine of absolute sovereignty' in the first case, stating that a state has absolute rights to water flowing through its territory, and 'prior appropriation' in the second, that is, 'first in time, first in right'.

These conflicting doctrines of geography and chronology clash along all of the rivers of the Middle East, with positions usually defined by relative riparian positions. Down-stream riparians, such as Iraq and Egypt, often receive less rainfall than their upstream neighbours and therefore have depended on river-water for much longer historically. As a consequence, modern 'rights-based' disputes often take the form of upstream riparians such as Ethiopia and

Turkey arguing in favour of the doctrine of absolute sovereignty, with downstream riparians taking the position of prior appropriation.

In many of the Middle East water disputes which have been resolved, however, the paradigms used for negotiations have not been 'rights-based,' either on relative geography or chronology of use, but rather 'needs-based.' In agreements between Egypt and Sudan signed in 1929 and in 1959, for example, allocations were arrived at on the basis of local needs, primarily of agriculture. Egypt argued for a greater share of the Nile because of its larger population and extensive irrigation works. Current allocations of 55,500 MCM/yr. for Egypt and 18,500 MCM/yr. for reflect these needs. Likewise along the Jordan River, the only water agreement ever negotiated (although not ratified), the Johnston Accord, emphasised the needs rather than the inherent rights of each of the riparians. Johnston's approach, based on a report performed under the direction of the Tennessee Valley Authority, was to estimate, without regard to political boundaries, the water needs for all irrigable land within the Jordan Valley basin which could be irrigated by gravity flow. National allocations were then based on these in-basin agricultural needs, with the understanding that each country could then use the water as it wished, including to divert it out-of-basin. This was not only an acceptable formula to the parties at the time, but it allowed for a break-through in negotiations when a land survey of Jordan concluded that its future water needs were lower than previously thought.

Because of its relative success, needs-based allocations have been advocated for the region in recent disputes as well, notably in and around the Jordan River watershed where riparian disputes exist not only along the river itself, but also over several shared groundwater aquifers. Shuval, for example, argues for a minimum baseline allocation between Israel, West Bank Palestinians, and Jordan, based on a per capita allotment of 100 m sup3 /yr. for domestic and industrial use plus 25 m sup3/yr. for agriculture. Adding 65% of urban uses for recycled wastewater, Shuval arrives at a hypothetical 2022 allocation as 950 MCM/yr. for Palestinians, 1,330 MCM/yr. for Jordan, and 1,900 MCM/yr. for Israel. Since the regional annual freshwater supply is only about 2,500 MCM/yr., Shuval also advocates a series of water import schemes and desalination plants to provide the difference between regional supply and future demand.

Wolf likewise advocates a needs-based approach, but considers new sources such as recycled wastewater as separate issues. By planning for total urban needs of 100 m sup3/yr. per person, and extrapolating to the point in the future where all of the basin's 2,500 MCM/yr. has to be allocated first to these needs, in other words when the regional population reaches 25 million, expected in the early part of the next century, Wolf arrives at allocations of 1,000 MCM/yr. for Israel, 1,000 MCM/yr. for Jordan, 300 for the Palestinians on the West Bank, and 200 for those in Gaza.

Although needs-based negotiations have been more successful in practice in the Middle East than rights-based claims, this is not to say agreement has been absolute. As noted above, agreement along the Nile has included only two of the nine riparian states Sudan and Ethiopia, both minor contributors to the river's flow. No other state riparian to the Nile has ever exercised a legal claim to the water allocated in the 1959 treaty. The notable exception, and the one which might argue most adamantly for greater sovereignty, is Ethiopia which contributes between 75-85% of the Nile's flow. Political complexities have likewise hindered ratification of water-sharing agreements along both the Jordan and the Euphrates Rivers, with upstream versus first-user arguments still being prevalent. Success has often depended on how well the negotiating strategy coincided with the political complexities of the region.

Economic Equity

One lately emerging principle incorporated into water conflict resolution is the allocation of water resources according to its economic value. The idea is that different uses and users of the water along a given water way may place differing values on the resource. Therefore, equitable water-sharing should take into consideration the possibility of increasing the overall efficiency of water utilisation by re-allocating the water according to these values. This principle alone may not be accepted as equitable by the parties involved. However, inclusion of economic aspects in water resource allocation may enhance better cooperation and future collaboration in joint projects in the region of concern.

Central Planning vs. Market Approaches

Allocation according to the economic value of water has usually been demonstrated using two approaches. The long-standing approach assumes a central planning authority who knows what is best for society a "social planner" who views the region as one planning unit. The social planner maximises regional welfare subject to all available water resources in the region and given all possible water utilising sectors. In some instances the social planner (government) also includes preferences (policy).

Second approach is the "water market" approach which employs the market mechanism to achieve an efficient allocation of scarce water resources among competing users.

Examples of these approaches can be found in several studies. Howe and Easter derived necessary conditions for economically efficient interbasin water transfer in the U.S Additional studies considered institutional and economic aspects of international cooperation for interbasin development. Goslin examined the economic, legal and technological aspects of the Colorado River Basin allocation between the US riparian states and Mexico. Krutillas analysed the economics of the Columbia River Agreement between the US and Canada. LeMarquand has developed a framework to analyse economic and political aspects of a water basin development. And Haynes and Whittington suggested a social planner solution for the entire Nile Basin.

Recent studies have questioned the equity and justice associated with market allocations. The conclusion from these studies is that economic considerations alone may not provide an acceptable solution to water allocation problems, especially to solve water allocation disputes between nations. While the social planner and the market approach may provide a unique solution to a problem of regional water allocation, they suffer several drawbacks that may affect the efficiency and the acceptability of the proposed solution. The social planner approach assumes that all social preferences are known and incorporated into the regional objective function. This of course might not be the case, especially when dealing with regional water allocations that involve many countries with cultural differences and preferences.

The market approach assumes the existence of many parties in the region, each acting independently, so that the market price for water reflects its true value for each party. If, in that market, one party's decision does not affect the outcome of other individuals, then the self-interest of the parties lead to an efficient outcome for the whole region. In the case of water, one party's decision may affect another party's outcome, creating what is called an externality or, third party effect. If the externality effect (cost) is not included in the supply curve of water, the market mechanism collapses. This introduces inefficiency into the system and results in what economists call "market failure." In the case of water (in a water basin), the externality effects might be multidirectional. This is particularly true for water basins shared by more than one country, and for water used for more than one purpose. Also, water allocation problems are not exactly similar to market setups that we are familiar with (e.g., the market for cars), because they are characterised by a relatively small number of agents with different objectives and water-related perspectives.

Game Theory

Game theory is an approach that allows the incorporation of economic and political aspects into a regional water sharing analysis with a relatively small number of participants, each with different objectives and perspectives. The principles of game theory are not discussed here in detail, but can be found elsewhere. For the game theorist, the dichotomy of whether two riparian states or political entities work unilaterally, in pursuit of only their own goals or, alternatively, work cooperatively in pursuit of regional goals, is recognisable as a familiar two-player, two-strategy game. In the language of the theory, unilateral water resources development might be referred to as a "defection" strategy while working together is referred to as a "cooperation" strategy. Each player chooses a strategy depending on such factors as regional geopolitical relations, relative levels of economic development, and riparian position.

For two water basins within the same political entity, with clear water rights and a strong government interest, the game may resemble a what is known as a "Stag Hunt," where mutual cooperation is the rational strategy. Between somewhat hostile players, either within a state but more often internationally, the game becomes a "Prisoner's Dilemma," where, in the absence of strong incentives to cooperate, each player's individual self-interest suggests defection as the rational approach. One example of this might be the Nile basin. In cases of high levels of hostility, game of "Chicken" can develop, with each player competing to divert or degrade the greatest amount of water before the opponent can do the same. The southern-most part of the Jordan River might be used as an example of riparians playing a game of "Chicken," with Syrian, Jordanian, and Israeli unilateral diversions all impeding basin-wide cooperation.

To cooperatively solve the problem of water allocations within the water basin, the parties involved should realise some mutual benefit that can be achieved only through cooperation. In cases of cooperation, each party needs to voluntarily participate, and accept the joint outcome from the cooperative project. Once a cooperative interest exists, the only problem which remains to be solved is the allocation of the associated joint costs or benefits. For a cooperative solution to be accepted by the parties involved, it is required that (a) the joint cost or benefit is partitioned such that each participant is better off compared to a non-cooperative outcome; (b) the partitioned cost or benefit to participants are preferred in the cooperative solution compared to sub-coalitions that include part of the potential participants, and (c) that all the cost or benefit is allocated.

The economic literature dealing with application of game theory solutions does not provide many examples of regional international water sharing problems. Rogers applied a game theory approach to the disputed Ganges-Brahmaputra sub basin that involves different uses of the water by India and Pakistan. The results suggest a range of strategies for cooperation between the two riparian nations which will result in significant benefits to each. In a recent paper, Rogers further discusses cooperative game theory approaches applied to water sharing in the Columbia basin between the US and Canada, the Ganges-Brahmaputra basin between Nepal, India and Bangladesh, and the Nile basin between Ethiopia, Sudan and Egypt. In depth analysis is conducted for the Ganges-Brahmaputra case where a joint solution where each country's welfare is better oft' compared to any non-cooperative solution.

Application of metagame theory, which is a nonnumeric method to analyse political (nonrational) conflicts, has been applied to water resources problems by Hipel et al. The resulting outcome of a conflict is a set of strategies most likely to occur and their payoffs to each participant. Becker and Easter have analysed water management problems in the Great Lakes Region between different US states and between the US and Canada. A central planning solution is compared to a game theory solution with the result being in favour of the game theory solution.

Dinar and Wolf, using a game theory approach, evaluate the idea of trading hydro-technology for inter-basin water transfers among neighbouring nations. They attempt to develop a broader, more realistic approach that addresses both the economic and political problems of the process. A conceptual framework for efficient allocation of water and hydro-technology between two potential cooperators provide the basis for trade of water against water saving technology. A game-theory model is then applied to a potential water trade in the western

Middle East, involving Egypt, Israel, the West Bank, and the Gaza Strip. The model allocates potential benefits from trade between the cooperators. Main findings are that economic merit exist for water transfer in the region, but political considerations may harm the process, if not block it. Part of the objection to regional water transfer might be due to unbalanced allocations of the regional gains, and part is due to regional considerations other than directly related to water transfer.

The history of hydropolitics along the rivers of the Middle East exemplifies both the worst and the best of relations over international water. While shared water resources have led to, and occasionally crossed, the brink of armed conflict, they have also been a catalyst to cooperation between otherwise hostile neighbours, albeit rarely and secretively. The major barrier to water's role as an agent of peaceful relations is the lack of a widely accepted measure for equitably dividing shared water resources. Many disciplines offer tentative or partial guidelines, including legal equity, needs-based equity, and economic equity. Each of these measures alone, however, cannot incorporate all of the physical, political, and economic characteristics which are unique to each of the world's international waterways.

In this chapter, we have tried to present the current state of equity measurements in water resources in general, and in Middle East water resources in particular. On-going research is attempting both to integrate each of the most useful theoretical approaches into one encompassing set of guidelines, and in quantifying the results for practical application to any water conflict. With the flow of water ignoring political boundaries, and with appropriate measures of water equity eluding disciplinary boundaries, the history of hydropolitics in the Middle East may give one a glimpse into a gloomy but probable future for many of the more than 200 international river basins. Without agreed-upon criteria for fair ownership and distribution of such a vital resource, many may come to experience the sentiments of Byron:

"Till taught by pain, men know not waters' worth."

The Valuation of Contaminated Land

The valuation of undeveloped land can be one of the most challenging appraisal assignments. One of the reasons can be the uncertainty of its highest and best use. Another is the uncertainty of the timing of its highest and best use, due to business cycles and other factors. The assignment becomes even more challenging when it is found that the land is seriously contaminated. The contamination might alter both the highest and best use of the land and the time required to obtain the regulatory approvals that are required to develop the site.

A client will often ask an appraiser to simply value the land under the assumption that it has no contamination. In other cases, the client may provide the appraiser with an estimate of the cost of remediating a contaminated site. The expected assumption is that the property be valued by simply subtracting the estimated remediation cost from the value of the acreage, assuming that it has no contamination. Both of these scenarios have occurred because, as Simons (1998) notes, there are few appraisers with training adequate to place a value on a contaminated property "as is".

The following is an example of an actual appraisal problem that was used to create a presentation for a study dealing with stigma. The property is referred to as Property A. A developer wanted to build a shopping centre on Property A, a site that cost $48 million. The purchaser paid $24 million in cash and financed the balance with a $24 million loan. Some contamination was found while soil borings were being done for the foundation. A remedial investigation was subsequently made. The resulting 'approved' cost of cleanup was estimated at $8 million.

Let's say that this $8 million cost is financed by a second trust deed. The appraiser for the transaction valued the property at $40 million, 'as-is,' or subject to the contamination. Many would say that a loss in value due to stigma would also occur. If you were the owner of the second mortgage and the owner of the first mortgage foreclosed, would you pay the $24 million required to protect your interest? If not, you have lost $8 million. You would be buying the property and its contamination for $24 million. How many of you would pay off the $24 million dollar mortgage?

A methodology is needed to quantify the downside risk of the $8 million remedial cost seriously understating what will eventually be the total remedial cost in order to answer questions like these. What about the risk of the land losing value whiles the remedial action is in process? It could well be that the land could lose its highest and best use during this time frame due to the development of other nearby retail sites

Lenders are not the only ones that need to have these risks quantified. Any purchaser of contaminated land needs to quantify the financial risk. There is also a need for accuracy for corporate balance sheets. If a major environmental liability is not recognised, the value of an asset can be significantly inflated. As Guntermann notes (1995), there is a significant amount of literature that documents reductions in property value around toxic, chemical and solid waste landfills and an emerging literature on stigma-related damage. The literature referenced in this study was confined to that dealing with the valuation of land that is itself contaminated, not other real estate that is near a contaminated site.

The seminal article by Patchin (1988) provided some examples of how he has valued contaminated property, including one example of how he valued contaminated land - the topic of this study. In the example, he included the cost of cleanup as a line item number and did not discuss the uncertainty of the estimate. The example also assumed that the land lost its highest and best use, since technological constraints precluded remediation to the point where its prior highest and best use would have been allowed. Patchin noted that when some appraisers found that a property was contaminated, they assumed that it was unmarketable and therefore that it was worthless. Patchin disagrees by saying, "the extent and nature of the contamination are the crucial factors in estimating the after value of a contaminated property."

In his second article, Patchin (1991) dealt specifically with the concept of stigma that he broadly defined as a loss in value beyond the cost to cure the contamination itself. He went on to list stigma as being caused by the following factors:

1. Fear of Hidden Cleanup Costs
2. The Trouble Factor
3. Fear of Public Liability
4. Lack of Mortgageability
5. Residential vs. Commercial
6. How Clean is Clean?

Patchin discussed the uncertainty in cleanup cost under the first item above and quotes Houston (1989) as saying that "If the estimated cleanup cost is 'X',the typical buyer will deduct '2X' from his purchase offer. Both Patchin and Wilson describe each contamination problem as being as unique as a fingerprint and Wilson (1992) stated that a consultant should provide a variety of costs. These are the most probable cost, the expected cost and a probable cost range. He gave an example of the cost range from a current assignment. Wilson quoted costs that ranged from a low of $5.5 million to a high of $63 million.

The second issue refers to the return that should be allocated to an entrepreneurial effort that is required to deal with the contamination. An allocation for entrepreneurial profit can be readily made by the use of a developmental approach to value. The third item refers to the possibility that soil contamination will seep onto another owner's property. The likelihood of this seepage happening can be assessed by the use of soil borings from a remedial investigation and can also be quantified.

In the discussion of the fourth factor, Patchin provides an example of a situation in which a property owner cannot get financing for new construction until the remediation has been completed. The prospective future value at the time of completion of the remediation is discounted for the five years of delay required to clean the site.

Healy and Healy (1992) and Kinnard (1996) surveyed lenders to find out how decisions on lending are affected by contamination issues. Both surveys found that one of the greatest concerns on the part of lenders was groundwater contamination. Kinnard devised a survey that shows 35.6% of lenders and 25% of investors avoid property with known soil contamination. The survey results also show that 45.8% of lenders and 41.1% of investors avoid property with known groundwater contamination. Soil contamination can lead to groundwater contamination in some situations.

With regard to the fifth factor, he states that market reactions to contamination differ according to whether the property is residential or commercial. And the sixth factor deals with the uncertainty of the level of cleanup that is required. The cleanup level is expressed as allowable levels of risk to the health of the nearby population, because of the contamination that remains in the soil after remediation.

Mundy (1992a) referenced Patchin's latter article and quotes seven criteria from another reference used to evaluate and determine the degree of stigma. Mundy says that stigma results from perceptions of uncertainty and risk. He noted that it may be relatively easy to quantify a simple contamination problem, but that as the complexity of the contamination increases, the level of uncertainty and perceived risk rises. Mundy then goes on to differentiate between real and perceived risk. Perceived risk is the risk envisioned by those who do not understand the true risk of the contamination.

Perceived risk could account for the differences that Patchin observed in stigma in residential vs. commercial properties. Very few homeowners are able to quantify environmental risk. Most buyers of commercial properties have environmental site assessments prepared. A buyer of a commercial property that did not have a study done to quantify the risk of contamination is not well informed.

Both Mundy (1992b) and Wilson (1994) discuss the variation in value (or lack of) at varying points in time. It is acknowledged that there may be no market for a contaminated property (and no value) until the seriousness and extent of the contamination is quantified. The contention is that a property increases in value while the site is being remediated and that the stigma may eventually disappear. Mundy (1992b) and Chalmers and Roehr (1993) focus on quantifying the loss in value of improved property by using cash flow analysis that provides for a remedial cost estimate over a ten year or so holding period. The discount rate that is used is supposed to reflect stigma by accounting for the "increased risk associated with the contaminated property," which the authors say is difficult to justify.

Lizieri, Scarlett and Finlay (1995) surveyed market participants in the United Kingdom in order to find out how they were dealing with the issue. They found a range of valuation methods that were used. The methods "range from their traditional all risks yield approaches and direct application of costs, through yield decomposition techniques to explicit discounted cash flow scenario modelling approaches and real option pricing models." They concluded that "the property industry is well able to identify problems, but is some way from establishing solutions."

Dixon (1995) concludes from his study of contamination issues in America that a systematic approach to valuation / appraisal of contaminated land is essential. Dixon refers to Wilson (1992) and his contention that a valuation "team" approach consisting of experts in such areas as appraisal and engineering be used and that an appraiser should be the manager of this team. Dixon also provides a quote from Campanella (1995) who sees the appraiser as a synthesizer of all relevant information, including remediation costs.

The appraisal professional is both the beginning and the end of the valuation process, providing the unimpaired value opinion from which all of the adjustments to value are made and providing the market place insight and technical value adjustment skills required to arrive at the impaired value opinion.

Weber (1996) agreed with Wilson and Campanella and described how the techniques used to quantify environmental risk to human or other biologically valuable receptors could also be used to quantify the financial risk of the contamination. This work described the remedial investigation and feasibility study that is required in the United States for a seriously contaminated site.

It also addressed the uncertainty that any one remedial method would be chosen and the Applicable and Relevant and Appropriate Requirements (ARARs) that govern which method is chosen. More recently, Chalmers (1996) notes that the quantification of risk in light of particular, site-specific factors is the most challenging part of the valuation problem. Chalmers states that risk analysis and quantification must begin with the perceptions of market participants. If the potential buyers of a site are well informed, perceived risk would equal real risk, as defined by his earlier article.

Jackson, Dobroski and Phillips (1997) note that all real estate has risk, but the type of risk that keeps real estate from selling is the risk that cannot be defined or quantified. They have concluded that sellers, lenders and buyers need to understand all of the risks in contaminated real estate and that it is necessary to quantify the risks associated with a specific property. He states that this risk is "always dependent on that property's unique environmental, locational, market, regulatory, and other characteristics." This study deals specifically with a methodology for quantifying the above-described risk.

A Thesis for Investigation

A number of consultants, e.g., Patchin (1988), Wilson (1992, 1994) and Roddewig (1996) have taken issue with the use of comparable sales for the valuation of contaminated land. This opinion is understandable since each site is unique relative to a number of attributes. References on hydrology and other soil sciences, such as Dragun (1988), Fetter (1994) and Freeze and Cherry (1979) show that, even if two sites otherwise identical (above grade) could be found with contamination, the type and extent of the contamination, the difference in subsurface conditions and proximity to an underground water supply could result in one site having one thousand times the environmental, and hence financial, risk of the other site. Travis and

Doty (1990) note that some sites may not be capable of remediation.

Since comparable sales are not likely to be useful in valuing sites with a significant contamination problem, a different valuation thesis will be investigated. The thesis is that a development model that would take into consideration the uncertainty of the remedial cost and prospective future value upon completion of the remedial effort would provide a more reliable estimate of value. The model would make use of Monte Carlo sampling to quantify the financial risk of successfully remediating a site. In order to illustrate this point, let us assume that three sites at three corners of the same intersection have all been contaminated. One from a gas station, a second from a dry cleaner and the third from a car repair business.

Assume that Property A, the site discussed in the Introduction, is located at the northwest corner of the hypothetical intersection of X and Y streets. Property B is located at the northeast corner; Property C, at the southwest corner of the same intersection. These sites are identical except for the type and extent of their contamination. Each has a value of $20,000,000 if not contaminated, as indicated by the recent sale of Property D, located at the southeast corner of the same intersection. The remedial costs of the three sites have been simulated, and the resulting frequency distribution has been overlaid on the locations of each property on the map.

Relatively common forms of continuous distributions have been overlaid upon the simulation of each remedial cost. The assumption implicit in the lender's appraisal was that the probable remedial cost for Property A was normally distributed, similar to that of Property C. The risk analysis shows that the distribution of remedial cost for Property A is much more skewed. It is similar to a Weibull or Gamma distribution. Property C is identical to the sites on the other three corners of X and Y, with the exception of the type and extent of contamination. C's contamination is all near the surface. It will cost, at most, $8.6 million to dig it all up, move it all elsewhere, and fill in the excavation.

A geographic information system (GIS) was used, to show that the contamination has seeped further into the soils at Property A. This GIS program can also calculate the environmental risk to humans from polluting the groundwater below the site-the only local source for drinking water. The minimal remediation cost is $8,000,000. If it does not work, a different action with a cost of $20,000,000 will be required. There is still a risk that a $43 million remedial cost might be required to remediate this site. How should the properties be valued?"

The Data for the Analysis of Property B

The prospects of incurring large financial losses led to the development of a method of quantifying the financial risk involved in acquiring a contaminated site. It applies to sites that have had a preliminary investigation, which concludes that contamination is likely. This financial risk stems from the possibility that a potential buyer of the site might have to pay for very expensive remediation.

The data for the analysis of hypothetical Property B came from actual cost estimates for another "test" property. The reversionary value in the example is also from the other property. The distribution of uncertain cost is the only item from the analysis that follows that is used to illustrate the risk in purchasing Property B.

Property B is representative of a number of sites that contain underground storage tanks and connecting lines for filling and dispensing material. The material could become a hazardous waste if it were to leak from its container. The financial risk stems from the likelihood that a leakage could have occurred anywhere in the underground system. This leakage could result from improperly pouring toxic chemicals down drains in the past. If so, the chemical could endanger the health of the surrounding population and would require removal.

The possible sources of contamination are described in terms of four units:

- Unit I the underground tank and filler/dispenser lines;
- Unit II an oil drain pit;
- Unit III a waste oil tank;
- Unit IV a clarifier.

An engineer performed a Phase II Environmental Site Assessment in order to confirm the presence of contaminants. It revealed oily residue in the surface soils over most of the site. Tests showed that metals and petroleum hydrocarbons were present at a variety of locations and depths. There were three locations where gasoline components were present and there was some hazardous waste in the clarifier unit.

The process requires that each unit be analysed to determine the risk of the buyer having to incur extensive cleanup cost. The risk analysis involved: breaking the in-ground construction into logical components; identifying major events; estimating the likelihood and cost of each event; and deriving a probability distribution of potential costs. The risk analysis goes beyond traditional single-point engineering estimates of the most likely, optimistic and pessimistic costs. Environmental engineers are surveyed. These engineers develop a range of possible outcomes from the remedial investigation and the cost of each outcome. The experts also provide the probability of incurring each cost. An event tree lists each of the units. The tree shows the possible remedial requirements and the resultant cost of each one.

The Conceptual Model

The solution to the problem was to structure it as a land development model. The reversion is an estimate of the prospective future value of the site. Starting with Unit I, there is a chance that remediation of the underground tank might be complete in step A by simple excavation. The reversion might occur at Time T=2, if step A in each section resolves all of the potential problems

Chances are that at least one of the units will require steps B or C, etc. If this happens, it will both add an additional "development cost" to the timeline and push the reversion further into the future. This is because regulators will not provide a letter of No Further Action (NFA) until all of the units are satisfactorily remediated.

The probability of each of the steps having to be taken quantifies the "Chances" referred to above. Multiply each cost by its probability to determine the subtotal for each cost. The next step adds the subtotal cost. The bottom line shows the total remedial costs for all of the units, followed by the reversion.

This model simulates the process thousands of times, collecting the present value that results from each iteration. The model takes all of the value estimates and creates a probability distribution of the value. The appraiser uses the range and probability distributions of the major costs that were provided by the engineers and assigns the likelihoods to the potential outcomes for the construction of the event trees. A detailed description of the simulation and its inferences may be noted by reference to the Addendum. A number of papers were presented on business cycles at the first meeting of the International Real Estate Society. Real estate researchers from Australia, Great Britain, Stockholm, and other parts of the world have noted declines in property values resulting from overbuilding as a result of the business cycle. Jeffries (1996), The Royal Institute of Chartered Surveyors (1995) and Born and Pyhrr (1995) noted the problems in assuming constant trends in values. The method described here can also be used separately from a valuation model for improved property that takes into consideration business cycles. In other words, a property could be valued by the use of business cycle analysis as described by Weber (1997) instead of having to rely on the Ellwood Technique for the income approach, which is recommended in a number of articles.

Again, for longer remediation periods, the risk of a differing highest and best use upon completion of remediation should also be considered for the valuation of undeveloped land. Fanning (1994) provides a number of reasons why this market analysis is so important in estimating future values. If another shopping centre had been developed within the trade area of Parcel A (as was actually planned), it could have resulted in Parcel A losing half its value from this factor alone. A probability distribution was also used to estimate the value of the property upon the completion of remediation. A triangular distribution (for the "test" property) with pessimistic, most likely and optimistic estimates were chosen.

Probabilistic Present Value Analysis

The results of each of the simulations of the remedial process are used in the spreadsheet for a present value analysis that quantifies the uncertainty of the future value. An explanation of the process follows.

Simulated Cost and Time for Remediation

This section contains a summary of the results of each spreadsheet simulation. The sum of the individual remedial costs for this iteration is labelled Total Remedial Cost.

Most Likely Value after Remediation

This reversionary value was derived from this iteration of the sampling of the triangular probability distribution that was used to estimate the value of the reversion after remediation. Estimates are made of the most likely, optimistic and pessimistic reversions for cash flow studies. A pessimistic scenario could reflect the loss in value in the event the site loses its highest and best use by the time that remediation has been completed.

Remediation Expense

Each estimated cost of remediation is spread out evenly over the time (in this case, quarters of a year) that is estimated to be required to complete the remediation of each Operable Unit. Again, the time estimate has been tailored to suit the specific outcome of the remedial sampling of the tasks to be done.

Real Estate Tax

This annual expense is estimated based upon the results of the most likely value estimate after remediation. It also has to be allocated over time based upon the periods that are chosen for the analysis. Other items such as entrepreneurial profit could be structured appropriately and deducted over time in this section of the model.

Cash Flow

The cash flow analysis discounts (at a 3% quarterly discount rate in this example) the cash flow after total remedial cost, real estate taxes, etc., and the reversionary value.

Estimated Value "As-Is"

This value indication is based upon the results of this iteration of the remedial process. The value indications for all 10,000 iterations of the spreadsheet have been plotted in the frequency distribution. The resulting mean value of the "test" property prior to the remediation amounts to about $1,700,000.

Calculating Stigma

The model can also be used to quantify stigma, if stigma is defined to include the financial risk of remediation. An appraiser can communicate the financial risk of a site as part of the process of confirming sales, once the risk of cost overrun is quantified by a probability distribution. Reference to the probability distribution of remedial cost for Property C shows that it will cost, at most, $8,600,000 to remediate this site due to the complete removal of the contamination. Survey research can be used to find out what level of risk potential buyers would take to remediate this site in light of the certainty of its profit potential if redeveloped. If, for example, it was found that the most likely buyers wanted a 95% certainty of covering remedial cost, it is a simple process to find the cost that will provide this level of certainty by finding the point on the X-axis that represents 95% of the area under the leftmost portion of the curve.

Let us say that $8,400,000 is the cost on the X-axis that corresponds to a 95% confidence level. Stigma would be reflected by the difference between $8,400,000 and the most likely remedial cost estimate of

$8,000,000. It amounts to $400,000, which is only 5% of unimpaired land value. Application of the risk analysis to Property B shows that mean remedial cost is just $35,000. A 95% confidence level for remedial cost here would only suggest stigma amounting to $20,000, or .1% of the value of a $20,000,000 site.

If the site were only worth $20,000 after remediation, the stigma would amount to 100% of the land value. This exemplifies why the calculation of stigma as a percentage of value is not considered to be an appropriate methodology. Probabilistic analysis can also be applied to Property A to determine how much the value of the property should be discounted over and above the $8 million "most likely" remedial cost in order to satisfy the maximal risk requirements of the market for this specific site. It may be that the probability of having to incur the $43,000,000 cost will be great enough to render the property valueless. At least the risk will be quantified so it can be more readily communicated to others. Potential purchasers can now be well informed about the financial risk of the site and perceived risk will equal real risk by the use of this methodology.

The methodology presented here was described to a speaker from an insurance company that provides insurance for contaminated sites. The speaker noted that the company estimators performed similar analysis in order to determine an appropriate premium when they consider providing insurance for seriously contaminated sites. The technique then passes one of the initial requirements of the Daubert factor, which calls for general acceptance of the technique, although it is no longer a precondition to admissibility. The technique could also be used to calculate stigma for an income-producing property, by quantifying the annual "premium" that should be deducted from income due to the financial risk of contamination.

The valuation of contaminated land can be among the most challenging appraisal assignments. One could argue that it requires that the appraiser have a number of atypical skills, such as a good understanding of environmental site assessment and remediation, the use of GIS systems, and significant experience in market and marketability analysis. Weber (1998) provides a number of examples of the use of GIS for dealing with environmental problems and for market analysis as part of the appraisal process. The literature suggests that appraisers are the logical ones to coordinate the efforts of other professionals.

Probability distributions can be a very effective method of communicating risk. An appraisal may be a waste of time and money if a remedial investigation that quantifies the risks involved has not been done. It also does not do an appraiser much good to simply know that a site has been contaminated and that somebody estimated that it would cost $8 million to remediate it. The implications of the cost simulations in the addendum are that percentage adjustments and risk rates are not appropriate for the quantification of stigma, if stigma is defined to include uncertain remedial costs. Comparable land sales with some form of contamination could be used as part of the valuation process, but they might only show one thing. This is that buyers do not refer to other land sales that also have contamination in order to decide what they will pay for a contaminated site.

A number of environmental engineers such as Swaroop (1987) have provided risk analyses, not unlike the one discussed in this study, to clients who are making purchase decisions. Textbooks such as the one by Covello and Merkhofer (1994) describe the use of Monte Carlo for environmental risk assessment. This study demonstrates that the same techniques can be used for financial risk assessment. The use of these techniques can result in a much better idea of the financial risk involved in the acquisition of contaminated land due to uncertain remedial costs and a better indication of its value.

Finally, Hoyt (1997) notes that the 1993 Supreme Court case known as Daubert has changed the admission requirements of scientific evidence by expert witnesses. He concludes that is imperative that real estate appraisers who testify be sure that their scientific evidence will stand up to the scrutiny of Daubert, or their testimony may be rejected. If there are tools such as the ones described in this study that are acknowledged to be useful in the quantification of environmental risk, it makes sense that appraisers should also use these tools in the valuation of contaminated land.

A Detailed Description of the Model

Investigations have to be made in order to get a letter from regulators stating that no further action (NFA) is required. Each remedial action for Unit IV and its corresponding probability of being required is as follows.

Event A: There is a 20% chance of getting by with only having to remove the clarifier and thereby receiving approval by the regulatory authorities.

Event B: There is an 80% chance of finding more serious contamination upon removing the clarifier. This would result in having to incur additional cost by having to do more testing. There is an 80% chance of getting an NFA letter after added testing. This is a joint probability, resulting in a .8 times .8 or 64% chance of not having to do more.

Event C: IF event B occurs after A (80% chance), there is a 20% chance of finding dichlorobenzene. If it is found, there is an 99% chance of NFA after an additional small remediation cost. This results in a .8 times .2 times .99, or 15.8% chance of incurring this expense.

Event D: This event assumes that the original testing (80% chance) and the added testing (20% chance) is required and that PCE is found (1%) but it has not migrated to the drinking water (99.9%), resulting in a probability of .8 times .2 times .999, or .1598%.

Event E: Event E is the scenario where everything goes wrong. The probability of PCE being found in the drinking water below the site is .8 times .2 times .1 times .001, or somewhat over one chance in one million.

Analysis of Units I-IV

This is the section of the spreadsheet that provides each estimate of the total remedial cost that might be required for closure of the site.

Column Headings

Remedial Uncertainties are the individual events that might be required to be done in order to attain closure. They are followed by the probabilities of the individual remedial tasks being required.

Joint Probabilities lists the probability that the individual task and all of its preceding tasks will be required. The joint probabilities are the ones that are incorporated in the model for sampling. For example, in order for event E under "Unit IV - Clarifier & Leach Field" to happen, it must be preceded by the failure of items A through D to solve the problem. The experts estimate that the chance of this happening is less than 2 in a million. Therefore, it is highly unlikely that the simulation will result in selecting this level of remediation as being required.

Remedial Cost is listed in the next column. It contains cost for each of the remedial tasks that might be required for each of the units.

Cumulative Remedial Cost lists the total remedial cost for each event A-E and all of the events that have to be completed prior to the sampled event. In other words, if the result of the sampling for "Operable Unit I - Tank & Lines" is that steps A through E have to be completed before a letter of NFA is granted, the total cost of the remediation would be $27,360. Expected Cost. This item is calculated by multiplying the Cumulative Remedial Cost by the probability of the cost having to be incurred. It is used to show the most likely cost of remediating the unit from a statistical standpoint.

Time to Closure: It can also be important to estimate how long it could take to remediate the site in the event significant contamination is found. Regulators may be present to test the soil when subsurface components are excavated. Certain types of contamination can take a very long time to remediate and building permits might not be allowed until remediation targets are met. The discounted present value of the land should be considered as part of a possible loss in value resulting from the contamination. The opportunity cost of carrying the land could result in a greater loss in value than the total cost of remediating the contamination. Cumulative Time lists the total time required to remediate the site, in the event more than one of the remedial steps are required. The cumulative time is used to estimate the value of the property at the time remediation has been completed.

Row Headings: Results of Sampling (Operable Unit IV as the Example)

Simulated Remedial Cost (IV). A custom probability distribution can be created to reflect the relative probabilities of each of the remedial uncertainties A-E. The distribution that was created for Unit IV is shown.

It shows the probabilities of outcomes A, B and C. The probabilities of Event D at .16% and Event E, with a likelihood of about 1 in 1,000,000, are so unlikely that they are only noticeable as dots on the X-axis. The statistical sampling process retrieves cost for spreadsheet analysis from cost distributions such as the above that are created for each of the Operable Units.

Simulated Time IV I Cost A-E. Contains the result of sampling the time function for the unit, given the number of steps (Costs) required (two, in this iteration).

Cumulative Cost at the bottom). Refers to the sampling results for the cost of each of the units (I-IV) required to successfully remediate the site.

Total Cost. This item refers to the summation of the sampled cumulative cost of each of the units.

Time. A vector containing the sampled time requirements for remediating each of the units.

Maximum Time. The simulated time required for remediation is the longest time required to remediate any of the units, based upon the outcome of a given iteration. In this iteration, the longest time that is required is 6.05 quarters for Unit III.

Simulation of Cumulative Remedial Cost

A computer-generated spreadsheet is used to simulate the remedial process 10,000 times for each of the units. The mean remedial cost ended up being about $35,000.

Implications of the Cost Simulation

The costs shown above were actual estimates for a contaminated property. They were used to help explain the process and to show that financial risk resulting from remedial uncertainty is relative. The resulting risk is significant if the property without contamination is just worth $50,000, but this risk would only have a nominal effect on the value of a $20 million site. The example shows why percentage adjustments applied to value or risk rates are not considered to be appropriate for quantifying stigma, if contaminated real estate is valued prior to remediation and stigma is defined to include uncertain remedial costs.

9

Hydrogeological Environmental Assessment

The safe and reliable long-term disposal of solid waste residues is an important component of integrated waste management. Historically, landfills have been the most common, environmentally and economically acceptable method of disposal of solid waste. Even with the implementation of waste reduction, recycling, and transformation technologies, disposal of solid waste in landfills remains a significant component of an integrated waste management strategy. Environmental impact assessment (EIA) can be defined as the systematic identification and evaluation of the potential impacts of proposed projects, plans, programs, or legislative actions relative to the physical, chemical, biological, cultural, and socio-economic components of the total environment. The prime purpose of the EIA process is to encourage the consideration of the environment in planning and decision-making to ultimately arrive at actions that are more compatible with the environment.

In India the environmental impact assessment exercises in the case of landfill projects are still at a development stage. Environmental legislation pertaining to the submission of an environmental impact assessment report for landfills was enacted two years ago. To date, about 50 sanitary landfill projects have been conceived, designed, and completed in India. But consideration of the environmental parameters in designing and developing these projects has been neglected. According to legislation on municipal solid waste-formulated and enacted by the Union Ministry of Environment and Forest as empowered under the Environment Protection Act of 1986 the submission of an environmental impact assessment prior to the designing and development of any landfill facility in the country has been made mandatory. Any municipal authority or designated agency engaged in the management of municipal solid waste anywhere in the country must perform an environmental impact assessment of the proposed site for a sanitary landfill operation. Landfill projects must comply with the standards for air, water (ground and surface), pollution, and other environmental norms.

The development of a sanitary landfill facility proposed by the local municipal corporation of the city of Jammu is justifiable as ensuring the establishment of a sound solid waste management system. Jammu has experienced massive urbanisation in recent years and the present system of solid waste management has failed to accommodate the quantity of solid waste that is presently being generated in the city. A preliminary environmental impact assessment study for the proposed project has the following objectives:

1. To compile an environmental inventory of the project area. This includes the following details:
   a. Hydrogeological settings
   b. Geographical and meteorological details
   c. Land use pattern of the study area
   d. Ground water quality of the study area
   e. Demographic details
2. To predict the possible impact on groundwater quality
3. To prepare an environmental management plan of the project site

Environmental Inventory of the Project Area

The study area of Jammu district, the winter capital of Jammu and Kashmir state, falls between 32°35'N to 32°55'N and 74°45'E to 75°00'E, covering an area of 870 square kilometres.

Physiography, Relief and Drainage

The study area is comprised of flat land and small hills. The main township is adjacent to lower Shivaliks, which are comprised of lower Himalayan Mountains. The topography is mostly undulating hills. The altitude in the area varies considerably from one location to another, the average height being 500 metres. The main drainage systems in the area are the Chenab, Tawi, Basantar and Ujh rivers.

The Chenab is perennial and snow fed. The rest of the rivers are seasonal rivulets known as Khads that traverse the area.

Climate and Rainfall

The climate of the area is subtropical characterised by three distinct seasons: winter, summer, and Monsoon. May and June are the hottest months while December and January are the coldest. The annual average ambient temperature varies from 2°C to 40°C.

Geology

The geology of the area consists of Shivaliks region systems and is mainly composed of sandstone and red transported quartzite. The lower accessible area, including the foothill plains, consists of alluvial deposits left by the seasonal rivulets, which carry away rainy water during Monsoon season. The parent material is mainly composed of alluvium and co-alluvium on the foothill plains. Accordingly, the study area was divided into recent, subrecent, and Shivaliks groups.

Soils

The soils of Jammu show complete heterogeneity. The soils of the foothill and adjoining area are composed of boulders and gravel with ferruginous clay. These types of soils are spread over in the study area and are generally loams, but poor in clay content. Soils on the foothills and V-shaped small valleys have been found to be deep with a medium to heavy texture.

Land Use Pattern

In 1995 around 35% of the region is under cultivation, 37% is forested, 29% is watershed, 0.25% is urban, and 2% of the area is snow cover.

Infiltration Characteristics

The soil infiltration rates vary under different conditions, land uses, and soil types in different hydro-climatic environments. Some recently conducted studies show that initial infiltration rates vary from 0.03 to 2.4 cm/hr for bare ground, 1.2 to 3.0 for agricultural land, 0.3 to 6.3 for grassland, and 0.6 to 1.2 for forestland respectively.

Demographic Details

The pressure on the available resources exerted by the population has an affect on the area. More than 1.2 million people live in Jammu, which comprises about 76% of the total area population; the rest of the population lives in villages. The population density is about 280 persons per square kilometre since the area is very easily accessible and is served by all modern means of transportation while the environmental conditions are conducive to settlement.

Existing Groundwater Quality

The average groundwater level in the Jammu region is at a depth of 18.8 metres below ground level. The average monitoring results of groundwater quality studies show that the general pH of the water varies from 6.5 to 8.5 showing its suitability for drinking purpose. The electrical conductivity values have been shown to exceed the limit of 750-2250 micromhos/ cm. Total dissolved solids have been found to be of 15800 mg/l. Calcium has been found to be within the acceptable limits, which is 250 ppm. The measured nitrate concentration exceeds the limit of 45 mg/l. The reason for these values can be attributed to agriculture runoffs, which result in transportation of nutrients from agricultural fields to water bodies. The fluoride was also found to exist in acceptable limits, which are 1.5 ppm.

Assessment of Hydrological Environment

Assessment of Aquifer Vulnerability using the DRASTIC Method

***Figure** : Hydrology, environmental protection agency]*

The prepared landfill project is supposed to cause some micro-scale impacts, which would be of significant nature on the soil or groundwater within the project area. These significant groundwater impacts are evaluated here using an aquifer vulnerability mapping technique called DRASTIC. DRASTIC is a methodology for identifying vulnerability to groundwater pollution. It uses seven parameters, which are a combination of geologic, hydrologic, geomorphologic, and meteorological factors, to relate an aquifer to the sources of its water and the constituents within that water. The parameters (Depth of water, annual Recharge, Aquifer media, Soil media, Topography, vadose zone Impact, and hydraulic Conductivity) are weighted according to their relative importance in determining the ability of a pollutant to reach an aquifer. The parameters are used to produce DRASTIC index numbers from which maps can be constructed.

The methodology was developed around a set of basic assumptions concerning a generic contaminant. They are:

1. Material introduced at the land surface as a soluble solid or liquid travels to the aquifer with recharge waters derived from precipitation.
2. The mobility of the contaminant is assumed to be equal to that of the groundwater.
3. Attenuation processes are assumed to go on in the soil, vadose zone and aquifer. DRASTIC was not specifically designed to deal with pollutants introduced in the shallow or deep subsurface, such as leaking underground storage tanks or deep mine wastes.

Ratings for each parameter are assigned depending on its contamination potential. Ratings vary from 1 to 10, with higher values describing greater pollution potential. These parameter ratings are used for all evaluations, however, each parameter is also weighted to account for differences between general land use and agricultural use. Weights range from 1 to 5, with higher weights representing greater pollution potential. This technique is based upon the seven parameters, or factors, of the DRASTIC rating scheme coupled with a relative importance weight resulting in a rating for each factor. The factors are:

D = Depth of groundwater

R = Recharge rate (net)

A = Aquifer media

S = Soil media

T = Topography (slope)

I = Impact of vadose zone

C = Transmissivity of aquifer (Conductivity)

Determination of the DRASTIC index number (pollution potential) for a given area involves multiplying each factor rating by its weight and adding together the resulting values. Higher sum values represent a greater potential for pollution or a greater vulnerability of the aquifer to contamination. For a particular area being evaluated, each factor is rated in a scale from 1 to 10 indicating the relative pollution potential of that factor for that area. Once each factor has been assigned a rating it is weighted. Weight values, from 1 to 5, express the relative importance of the factors with respect to each other. Finally the total impact factor score, the DRASTIC index number, can be calculated:

$$\text{Pollution potential} = D_rD_w + R_rR_w + A_rA_w + S_rS_w + T_rT_w + I_rI_w + C_rC_w$$

where

r = Rating for area being evaluated (1-10)

w = Importance weight for the factor (1-5)

Factor ratings are derived from data on each factor while importance weights are found in generic DRASTIC tables that list weights for factors having greater applicability (Aller, et al., 1987). The values of the factors for this study area were calculated from data from the southwestern Himalayan regional centre of the Indian Institute of Hydrology, Roorkee, India.

The data of the various parameters of the study area are as follows:

Depth of groundwater = 32.80 feet

Rate of net recharging = 9 in/hr

Aquifer media = sandstone with pebbles

Soil media = sandy loam

Topography = 5%

Impact of vadose zone = sand and gravel

Transmissivity of aquifer = 1729 pd/ft

This information was calculated into specific values for each factor:

Depth of groundwater = 5

Net recharge rate = 8

Aquifer media = 7.5

Soil media = 6

Topography = 9

Impact of vadose zone = 7.5

Transmissivity of aquifer = 8

These rating values are multiplied by their importance weight then added together to arrive at the DRASTIC index number:

Pollution potential

= DrDw + RrRw + ArAw + SrSw +TrTw +IrIw +CrCw

= 5 x 5 + 8 x 4 + 7.5 x 3 + 6 x 2 + 9 x 1 + 7.5 x 5 + 8 x 3

DRASTIC index number = 162

The DRASTIC index number (pollution potential) is high, thus showing the vulnerability of the aquifer to contamination from the proposed landfill project.

Environmental Management Plan for the Project

The environmental management plan for the landfill facility may primarily address the groundwater contamination and leachate management due to the terrestrial nature of environmental degradation caused by the project. The prevention of groundwater contamination would require the use of geosynthetic clay liners or membranes for the landfill. The contamination probability index calculated by the DRASTIC model was included in the remedial plan for the project. The high saturation of groundwater and the presence of alluvial soil media are the prime reasons that the project poses a contamination threat to the groundwater. Utilising the liner technology of geosynthetic clay materials can mitigate this threat. It was decided to put in a drainage system in order to provide swift and smooth removal of water from rainstorms. The geosynthetic clay liner is provided in sandwiched layers of cement and other stabilisation material. The groundwater monitoring will be done on a permanent basis in order to report any contamination caused by excessive leaching of heavy metals or other chemicals. Besides that leachate and landfill gas monitoring system, management of leachate and minimisation of soil erosion may include some aspects of environmental management of the landfill project.

The present environmental impact assessment of the landfill has been carried out to prepare a database of the impacts that could result from the project. In the present study, the environmental impacts that may occur are analysed with particular emphasis on groundwater contamination. This study has been able to identify the critical deleterious environmental degradation that could occur from the project. The database prepared for the environmental inventory of the project may be useful for other studies in the area regarding the landfill management plan and its implementation during the later stages of the project.

Water Pricing

The claim is often made that more effective use of pricing and various water charges can remedy or completely solve problems of water scarcity, shortage or overuse. A common view is that users pay too little for water and, as a result, they use too much. The public or private agencies that supply water to users could levy higher prices willingly, or the regulatory agencies that manage and oversee water resources could enforce higher rates. Implementing these approaches in developing countries may present special challenges due to the high costs involved or due to inadequate resources to assign, monitor and enforce diverse rights to water use. However, are there reasons that other countries, not so constrained, have not implemented effective water pricing strategies? Is this pricing prescription really so straightforward?

Economists, engineers and others have played a lead role in these policy debates, and economists continue to write volumes about many aspects of the issue. The purpose of this chapter is to review current approaches to effective water pricing and to examine the gap between these academic prescriptions and the current state of water allocation policy and practice. To what extent are more complex or comprehensive systems of water charges likely to fulfill the growing challenges of water scarcity? What steps are needed to bring theory into practice?

The main conclusions reached here are that in many jurisdictions, schemes of administered prices may represent an attractive policy approach to promoting the overall efficiency of water use. The attainment of appropriate or efficient usage levels is synonymous with encouraging conservation and reuse, with investment and innovation in new technologies and practices and with achieving expected levels of water quality and security of supply. It is also consistent with capturing as fully as possible the many beneficial consumptive uses of water in situations where water is relatively abundant or inexpensive. Unlike the allocations that might be reached in some private market transactions for water use, an advantage of administered pricing is that it offers the ability to include, as part of the prices charged, society's best estimates of social costs related to downstream or future users, instream flow uses, and so on. Unlike private market outcomes, administered pricing also offers the ability to make the water pricing program revenue-neutral. A carefully designed rebate program can leave the average water user's annual household or business income unchanged, even while providing strong financial incentives to conserve.

These issues and arguments are explored in the following sections. The discussion starts by examining the concept of water pricing and its rationale, focusing attention on those pricing approaches that promote the efficient use of water resources. The concluding sections highlight pricing issues that require further attention. Some brief illustrations of the state of water pricing in Canada and elsewhere show that despite initial steps toward policy reforms, considerable action is still required to harness the potential gains of effective water pricing.

The Meaning of Water Pricing

Various systems of water rates, water fees and user charges are employed around the world as a means of influencing the processes by which water is provided and used, and as a means of cost recovery or revenue generation. In some places, these water prices are the principal form of water allocation. In other places, these prices are combined with various forms of licenses, permits, quotas, restrictions and other practices and customs that dictate how much water is used, where and at what cost. Where pricing is used, there may be a range of apparent price levels since, in practice, water systems have a number of stages in the supply chain. There might be prices assigned for bulk water withdrawals from a surface water or groundwater source that differ from other prices further along the supply chain. Specific prices might be assigned once the water has been successively transported, stored, treated and distributed for final use by residents, industries, irrigators, public works and so on, and these prices may vary by time, place or purpose of use.

Economists have occupied themselves for centuries exploring the role and behaviour of prices in influencing decisions about the production and use of all manner of goods and services. Long before economists became involved, people had experience with barter and trade in town and village markets everywhere. It was readily apparent that markets could function with varying degrees of effectiveness for a wide range of transactions. At their best, markets and the price signals they generate are capable of coordinating the independent decisions of vast numbers of producers and traders and of allocating scarce supplies of goods and services across diverse consumer groups. It is also the case that markets work more effectively at allocating some types of goods and services than others. Historically, markets have not been the principal means of allocating the use of fresh water in most parts of the world.

Since, in most countries, an individual's ability to use groundwater or surface water is not decided by the interaction of numerous independent buyers and sellers, there is a potentially important distinction to be made between market prices and administered prices. Market prices are those that arise from the changing balance of supply and demand among competing traders, whereas administered prices are those that are set or imposed by some public agency in the absence of an interactive market process. The two types of price determination occupy opposing ends of a continuum, where the determining difference is the degree to which information about buyers' and sellers' valuations drive the price-setting process.

To illustrate this distinction, consider a regional water authority that acts as a single supplier and that sets prices based only on notions of its own cost of service. This pricing behaviour would place it at the administered pricing end of the spectrum, since users' valuations are not reflected in the prices charged. If the water authority's policies and strategies were to change, and the water authority chose to set prices reflective of buyers' willingness to pay, then the resulting market prices might be indistinguishable from those a profit-oriented monopolist also would have charged. Indeed, a central notion for the improvement of water pricing practices around the world is to shift or reorient price-setting objectives and/or processes at the local level. Especially where notions of market efficiency and the best use of scarce resources are paramount, the challenge is to make administered prices more closely resemble the market prices that a well-functioning market might set.

When prices are determined by auction or due to the competitive supply to the market by numerous sellers, an anticipated shortage is usually resolved spontaneously through a price increase. Alternatively, under the administered price approach, a conscious effort would have to be made to adjust prices in times of scarcity. If that effort were insufficient, then other measures such as quotas and water restrictions would be needed if all customers were to be assured some supply. In the face of an anticipated shortage, both pricing approaches do better if there is timely and accurate market information and forecasting. In the case of markets, any adjustment of prices in response to new forecasts is the implicit responsibility of all buyers and sellers collectively. Under the administered approach, the selling agency is the only one that can raise or lower prices. Under both pricing approaches, there remains a risk of shortage. Some of the cost of this risk and of any resulting shortage will be borne by both buyers and sellers.

Discussion about the design and reform of water pricing policy, then, is almost always about the purpose and relative effectiveness of various schemes of administered pricing as a means of allocating usage of water or raising revenue. There may or may not also be a policy option to establish some form of water market. With active water markets, there would be no specific role for pricing policies, since then the specific price levels are not a policy choice but the result of market interactions. Where there is a choice between new or reformed water markets and reformed water pricing, a clear understanding of reformed water pricing would help inform such a decision.

The Rationale for Water Pricing

Setting a price for water use can potentially achieve a number of diverse purposes, but an acknowledged limitation is that each price can generally only achieve one of these purposes and so priorities need to be established.

Typically, the three primary and competing purposes to be prioritised are promoting economic efficiency, generating revenue and advancing economic equity or fairness. Where a water pricing method includes two price elements such as a flat monthly charge plus a volumetric charge based on metred use then it may be possible to find specific combinations of the monthly charge and of the unit charge that can achieve two of these purposes simultaneously. A decision to focus on one prioritised purpose means that other policy instruments are likely to be needed to address the secondary purposes, or else they might go unfulfilled. A decision to not focus on or priorities a single objective could be consistent with a compromise pricing scheme that tries to satisfy all of the objectives somewhat but that is unlikely to achieve any of them fully.

Economic efficiency refers to a water allocation outcome in which nobody could be made better off such as through some other water allocation or water price accompanied by compensating transfers without making someone else worse off. Importantly, this efficiency goal the so-called Potential Pareto Efficiency comprises a number of desirable features or subgoals, including notions of water conservation; water quality improvement; encouraging innovation, investment and risk-taking; and ensuring security of water supply. That is, pricing that seeks to achieve economic efficiency seeks to frame the water allocation challenge quite broadly and to include all of these dimensions of water use that provide value or cost to society. For example, an administered water price that is set too low might encourage too much current consumption and not provide adequate incentives to preserve water for use in later periods or downstream. This would not be an efficient outcome if a higher price could save water and if the gains from subsequent use exceed the near-term sacrifices made to achieve them.

The competing goal of revenue generation may take the form of encouraging a public utility to operate on a not-for-profit basis or of allowing a privately owned utility to earn a fair rate of return on its investment. In other cases, revenue generation from water resources may contribute significant revenues to the public treasury and act as a "tax-by-another-name." Too little revenue generation may result in water use decisions that feature relatively low investment in technology, infrastructure and equipment. Too much revenue generation may result in high prices with too little beneficial use of the water that is available.

The potential mismatch between revenue raising and economic efficiency objectives in setting water prices has generated two main responses among economists. One approach favours choosing economic efficiency as the primary goal and using other public policies, including other taxes or subsidies to meet revenue targets, as needed. The other approach proposes multipart pricing structures more complex systems of water pricing that might be able to achieve both goals simultaneously. Authors in both groups agree that, in some jurisdictions, the historical pursuit of revenue generation objectives as a higher policy priority than efficiency objectives constitutes a barrier to solving water allocation problems that occur there.

The third competing goal of water pricing is economic fairness. Most of the attention here is on ensuring affordable access to some threshold level of water by all users, especially those with low incomes. The other strand of this discussion also seeks to promote fairness by placing high charges or surcharges on consumers of large volumes, under the presumption that such users are relatively wealthy and should pay more. Here there are also two main responses taken in the water pricing literature. One view is that, in both developed and developing country examples, the status quo is unwarranted. Specifically, low-priced or unpriced water is providing a greater benefit to relatively high-income households than to low-income households. Under such pricing schemes, wealthy users often gain access to more abundant, reliable supplies at a relatively low cost. Poorer users are often under-serviced and may go without water, use less water or pay more overall. In developing countries, the cost to the poor includes travel time to distant sources and high unit prices for purchased water not reliably available from the low-priced public supply. In developed countries, a low unit charge is sometimes accomplished by also charging a (correspondingly higher) monthly flat rate, and the overall effect disadvantages the poor.

The second response to this focus on fairness is to note that, especially in developed countries, the annual cost of threshold amounts of water use even under higher pricing that promotes economic efficiency would be a relatively small part of a low-income household's budget. For instance, these water payments would often be less than what is paid for access to energy, transportation, housing or entertainment. Where ability to pay is an issue, it is best to address it through income support measures and safety net programs, not through the de-linking of water costs from water use.

How should this controversy over the primary intended purpose of water pricing be resolved? Especially in jurisdictions where there are significant current or anticipated problems of water overuse, scarcity or quality, it seems likely that the value of water pricing as an instrument of public policy will yield its highest social return if it is directed toward the pursuit of economic efficiency. In this situation, goals for revenue generation and fairness are more likely to be met using available policy instruments other than water prices. Of course, the primacy of the economic efficiency goal and the lower relative priority and advantage of the competing goals will vary by jurisdiction, and could be compared using standard economic methods for evaluating policy choices. Sometimes, there will be multiple possible schemes to price water efficiently, such as when considering whether to supplement a per unit charge for water with a fixed monthly fee. In these cases, revenues and fairness need not be ignored. In general, decisionmakers considering a pricing policy should be careful not to jeopardize efficiency by saddling their program with revenue-raising targets as well.

In those cases where efficiency is the main goal of water pricing, there may also be other criteria that influence how and whether pricing reforms are implemented. These include policy flexibility, administrative feasibility, political acceptability and overall effectiveness. These criteria can play a role in the design and implementation of new reforms and in the ongoing assessment of established pricing systems. Historically, in some jurisdictions, water rates have rarely been adjusted, yet fluctuating water supplies and demands may recommend a flexible price regime accompanied by timely information dissemination about current prices to individual users. The flexible regime would make water pricing more closely resemble the way that some households and businesses contract for electricity and other energy in fluctuating markets. Seasonal pricing or more frequent rate adjustments would allow for users to adjust their own water uses according to its current scarcity value at any point in time. Relatively complicated pricing systems, such as those that vary by volume, time and point of use, require greater administrative capacity, including investments in monitoring or smart metering. In some jurisdictions, these services will be too costly or are not available.

Types of Pricing Promote Efficient Water Use

There seems to be agreement among economists about the principles for rate setting, even if these practices are not yet widely implemented. Without having to address concerns about revenue generation or fairness, the role of prices is to serve as an incentive for users and suppliers to reach quantities of use that are efficient. That quantity of use is one for which the value to the user of the last unit of water used equals the costs of supplying it incurred by that user and everyone else the so-called social marginal cost of supply. This cost will vary according to the source of the water and the time and place it is supplied, among other factors.

There is one principal advantage of administered water pricing compared to allocating water entirely through markets, auctions and tradable entitlement schemes. With administered prices, there is the option to employ a more comprehensive and relevant measure of supply costs than those reflected in most markets. For example, suppose users withdraw surface water from a river system. One should look beyond their costs of pumping or diversion to define a marginal water cost that also includes costs incurred by downstream users who will access less water. Suppose users withdraw groundwater from aquifers with storage capacity. In defining these social costs, one should ask not only about opportunities foregone by other current users, but also about costs to future users who will access less water. Two non-market valuation methodologies provide an excellent starting point for estimating such values. So-called stated preference and revealed preference methods can generate these estimates for incorporation into appropriate definitions of social marginal cost and administered water prices.

Attempts to reach the optimal levels of water use will require estimates of how these costs vary over space and time. In almost all parts of the world, there are relatively predictable patterns of water flow and use that depend upon climate and hydrology, as well as upon established cycles and patterns of human activity. Consider the situation in which there are one or more wetter months or seasons, and one or more relatively water-scarce months or seasons. To achieve efficient outcomes in seasons of water abundance, one should expect to see a lower price of water that reflects the lower marginal costs of providing it. Since this water is relatively abundant, the set of relevant costs might include only the costs of pumping, distribution and treatment, if any, but relatively little cost in terms of displacing or crowding out the demands of others (e.g., downstream or instream users). Conversely, in dry seasons, the relevant costs include the same wet-season elements plus all the associated costs of storage and of other system expansion that allow these dry season users to be provisioned with the expected degree of reliability and water security. The result will be a pattern of administered water prices that could vary significantly from month to month, for example. Under such prices, users would face strong incentives to conserve and to innovate in water saving, including optimal investments in methods and techniques that conserve or reuse dry season water. Similarly, users should know that when water is abundant, they will be allowed full access at low or zero prices, provided those prices cover the relatively lower costs of their current use in those months.

These pricing principles describe an approach to reaching efficient usage levels of water in times of both shortage and abundance. Even though these are administered prices, they are capable of providing important signals across the economy about preferred patterns of water use and about alternative means of supplying or provisioning water. For example, Grafton and Kompas examine the way that

Sydney, Australia, has coped with periodic droughts and water shortages without using water pricing effectively. When there is a shortage, Sydney authorities impose various schemes of water use restrictions. These restrictions impose costs indiscriminately on broad groups of residents and water users. Historically, there has not been much public interest in investing in permanent or standby water capacity to address periodic droughts and shortages. According to Grafton and Kompas, Sydney should impose prices sufficiently high to ration available supply during these drought episodes. The authors' estimates show that, at these prices, it would be immediately apparent that investments in alternative standby sources, such as desalination, would be warranted, even if the facility is expected to be used infrequently.

Notice that in no part of this discussion of pricing to promote efficient water use are systems of increasing or decreasing block rates involved, under which neighbouring users pay different rates according to their levels of monthly use. Indeed, if the cost of supply is the same, then efficiency is hampered if high block prices are arbitrarily curbing the water use of one household while allowing the household next door to consume at lower price levels. Whatever their other merits, such as for revenue raising or the pursuit of fairness, there seems to be agreement that these rate design features do not promote efficient water usage levels.

Higher water prices that are introduced to support efficiency goals might, in practice, generate significant revenues. This revenue collection is a side effect that is itself not necessary for higher or lower prices to guide users to efficient allocation decisions. If revenue generation is unpopular, authorities can exercise their option to implement revenue-neutral water pricing. Under such a scheme, authorities commit to returning all revenues to the water consuming population, provided that the rebate amounts are not directly related to actual water use. For example, the rebates could be used to lower property taxes or to eliminate the monthly flat rate portion of water and other utility bills.

Although there seems to be wide agreement about how these principles should apply to the pricing of water at both the wholesale and retail (residential) levels, some aspects of water pricing may not be well understood. Three issues that deserve brief mention here are differences between surface water and groundwater, water and sanitation linkages and the concept of user cost. The foregoing prescription for water pricing is to adopt a broad definition of marginal cost incurred by society in using incremental units of water. These marginal values should differ according to whether the source of the water is groundwater, surface water or some combination of the two. It is important not to overlook these cost differences, such as by using some average price or cost. Increasingly, water resource managers have come to understand that groundwater aquifers can play an important storage and conveyance role, and that groundwater may have extra value as a supply of last resort if surface water should become unavailable. Accurate price signals that reflect these differences can help to coordinate optimal systems of conjunctive water use. That is, the optimal response to a higher relative price for the groundwater could include using groundwater mainly in times of surface water scarcity and ensuring full groundwater recharge in times of surface water abundance. Further, all customers could gain from having serviced access to water from both sources, allowing customers to benefit from whichever is cheaper.

Especially at the residential level, there is a strong linkage between household water usage and household sewer and sanitation throughput. Increasingly, water usage can be metred and billed at low cost, but sanitation throughput cannot. There are some differences across households according to exterior (lawn and garden) water usage that affect the relation between the two service flows, but these can be estimated fairly directly. An issue for many jurisdictions is that sewage treatment, particularly at the tertiary stage when phosphates are removed, can be quite costly. Its cost may not be well understood or appreciated by water-using households and thus not reflected in households' consumption behaviour or choice of appliances, for example. Authorities can adjust their use of metred water pricing principles to reflect the complementary provision of sewage treatment services at the household or business level. The effect will be to increase the overall efficiency of the two services combined, likely reducing water usage levels in the process.

Water pricing practices and principles may not have paid enough attention historically to the cost of raw water, whether provided from groundwater or surface water sources. Raw water refers to untreated water drawn directly from natural water sources. Many water pricing regimes that focus on only cost recovery do not see raw water as a cost and do not charge for it. If these agencies adopt pricing principles based on the long-run marginal cost of water, they might still overlook it. Economists argue that, just like the other costs incurred along the way, one must include any cost incurred from using this "free" water now. Economists refer to these costs as marginal user costs. In the case of storable groundwater, the cost of using some of the water now is that water might become more costly to pump from greater depth or to treat for poorer quality in the future. In the case of surface water, using more of the water now might take away valuable reserve or standby capacity, necessitating the expense of developing other storages or supplies sooner than would otherwise be the case. Either way, these are costs to estimate and to include as part of an effective pricing regime. Given the accumulated writings and debates on the merits of alternative water pricing schemes, to what extent does one see evidence of their adoption already in practice?

Water Pricing in Canada and Elsewhere

Despite Canada's abundant freshwater, dais water is not always available when and where it is needed. Approximately 60 percent of the country's freshwater drains to the north, away from the 85 percent of the population living near the country's southern border. Due to economic growth, the demand for water in Canada has been increasing rapidly, posing a major challenge for water use and management. In some parts of Canada, water resources have become increasingly scarce and a constraint to the further expansion of those economic activities that heavily rely on water.

There have also been concerns about water quality stemming from contamination by municipal, agricultural and industrial activities. Poor water quality poses a threat to human health and increases the costs of using it productively Water resource managers now face concerns about the impacts that global climate change could have on water resources in some regions.

In Canada, the federal and provincial levels of government share the jurisdiction for water management, with some delegation by provinces to regional and municipal governments. Water prices differ widely across provinces and across municipalities within each province. At the provincial level, systems of permits regulate water use. Typically, there are prices charged for these permits to withdraw water, but additional permits may not be obtainable in many cases. Water allocations are more likely to be restricted by the quantity provisions of these permits than by the permit fees that are charged. Seven of the thirteen provinces and territories charge some type of one-time or recurring fee for permits to withdraw water. In the remaining provinces, there is no charge at all for them. These fees charged vary from $0.01 to $143 per 1,000 cubic metres of permitted annual water withdrawals.

The concerns about water resources in Canada are exemplified by the recent initiatives that the provinces of Alberta and Ontario have undertaken to improve water management. In Alberta, the govermnent's Water for Life strategy aims to provide safe drinking water, to maintain and protect aquatic ecosystems and to manage water in ways that support sustainable economic development. Part of this policy framework includes a commitment to examine more effective uses of economic instruments for water management, potentially including water pricing.

In Ontario, concerns about water use have motivated the introduction of a new water charge for industrial and commercial water users. Commercial and industrial water users who withdraw more than 50 cubic metres per day from groundwater or surface water sources or from a municipal system will be required to pay a charge of $.00371 per cubic metre of water starting in 2009. The money collected from the charge will cover a portion of the provincial costs of water management activities.

In addition to the fees levied directly by provinces, there is a remarkable diversity of water rates and water rate structures charged by regional and municipal water authorities, such as for water use by residents and businesses.

The main categories of rate structures are flat rates, constant rates, decreasing block rates and increasing block rates. It is estimated that 37 percent of households in Canada face a flat rate charging system (i.e., they pay a fixed monthly fee that is independent of the amount of water they use) and 39 percent of households face constant water rates (i.e., the unit price does not vary with the amount consumed). A further 13 percent of households face water rates where the unit price decreases with the amount consumed, and 10 percent face water rates where the unit price increases with the amount of water consumed. Note that for more than one-third of Canadian households with municipal water service, payments for water service are not in any way connected to the amount of water consumed.

Current systems of water pricing have come under criticism for diverse reasons that relate to the range of roles that water pricing can potentially play: efficiency, revenue generation and fairness. With respect to efficiency, there are concerns about the public's inability to obtain new water permits in regions where water is scarce and complaints about the general lack of effective water conservation measures or incentives. With respect to revenue generation, much concern is expressed when, in many cases, the revenues generated by water prices are insufficient to meet all the operating and capital needs of water suppliers.

One study estimated that only about 50 percent of these costs were being met. There are concerns about the fairness of some historical allocative provisions, such as those that allow holders of senior appropriative water rights to consume vast quantities of water without any payment, even under severe drought conditions. There are public expectations that some new water pricing system should solve all of these problems, yet this is simply not feasible. The approach suggested here is to focus pricing incentives on issues of water efficiency, and to address concerns about utility finances and about fairness through other available policy actions. Fortunately, price increases motivated by efficiency concerns may also help with revenue targets and fairness concerns, but are unlikely to resolve them completely.

Canada is not alone in expressing discontent about a wide range of water pricing structures and practices. As Young and McColl report, Australia is also struggling to find the appropriate water pricing scheme, even in the face of the nation's pervasive water scarcity and drought. Commenting on a recent survey by Australia's National Water Commission, Young and McColl describe the water pricing practices of fifty-seven urban water suppliers. Of these, twenty-five charge a flat service fee plus an inclining block rate (volumetric charge), twenty-four charge a flat service fee plus a constant volumetric charge, four use an inclining block rate without the flat service charge, one uses a declining block rate, and only three apply the flat service charge independent of the water volume used. As with the examples from Canada, if there is a pattern or principle to these varying pricing practices, it is not apparent.

A recent economic commentary on Australia's water woes concludes that there is a "long-term mismatch between demand and supply that can only be resolved by an increase in prices." In the European Union, water pricing has come to be regarded as an important tool that can be used to improve the management of water resources. The European Union Water Framework Directive (WFD) came into force in 2000 and aims to make the management of European water resources more efficient. The WFD aims to create new tools for sustainable water use, including measures for managing water demand and patterns of water use. Article 9 of the WFD obliges member states to take into account the principle of recovery of the costs of water services including environmental and resource costs. Article 9 further specifies that members have to ensure by 2010 that water pricing policy is an incentive for efficient water use and thereby contributes to the environmental objectives, and that different water uses make an adequate contribution to the recovery of the cost of water services. This approach to water pricing that the WFD mandates has considerable potential to resolve current inefficiencies in water management and to bring about a water use regime that is sustainable.

Despite the WFD water pricing principles, the actual water pricing practices followed in the member states differ considerably. So far there has been a mixed experience with the integration of economic and environmental objectives into the pricing policies of the member states. Overall, full recovery of financial costs is only partly achieved, and the environmental and resource costs are rarely considered. Water pricing policy faces its greatest challenges in the agricultural sector, especially in southern European countries, where agriculture is by far the largest consumer of water and where water scarcity problems are the greatest. In the European Union, as elsewhere, pricing is not the sole policy instrument that can or will solve numerous water resource problems, but it would appear its full potential has yet to be realised.

In many jurisdictions, schemes of administered prices are likely to represent an attractive policy approach to promoting the overall efficiency of water use. The attainment of appropriate or efficient usage levels is synonymous with encouraging conservation and reuse, with investment and innovation in new technologies and practices and with achieving expected levels of quality and security. It is also consistent with capturing, as fully as possible, many beneficial consumptive uses in situations where water is relatively abundant or inexpensive.

Unlike the allocations that might be reached in some private market transactions for water use, an advantage of administered pricing is that it offers the ability to include, as part of the prices charged, society's best estimates of social costs related to downstream or future users, instream flow uses, and so on. Unlike private market outcomes, administered pricing also offers the ability to make the water-pricing program revenue-neutral, such as by returning some or all of the fees collected to the water consuming population in any manner not directly related to actual water use.

10

Laboratory Exercise to Illustrate Pollutant Hydrology in Groundwater Systems

Cleanup of polluted groundwater sites is a primary goal of EPA s superfund budget. While pollutant transport is extensively covered in groundwater modelling lectures, it is completely absent in published laboratory exercises. This is true of most experiments involving soils or sediments, as noted by Tran et al. (2001), who found that of the 92 environmental articles published in the Journal of Chemical Education from 1969 to 2000, only approximately 15 percent pertained to soil matrices. Iran et al. (2001) further notes the persistence of a "perception that soil chemistry is more complex and challenging than water chemistry". Indeed, the relative lack of simple soil experiments may result from the variety of complex chemical and physical properties that can be present in soil and sediment samples (i.e. mineralogy, surficial coatings, organic matter, pH values, and EH values).

Even ifan educator develops and publishes an excellent soil experiment, others who follow the published procedure, inevitably with different soils, will likely ootain widely varying results. Topics covered by the few relevant environmental experiments from the literature that involve soil matrices include various sorption reactions, soil remediation projects, and the interaction of acid rain with soil weathering. We describe two new laboratory experiments using soil matrices in our manual on instrumental and environmental chemistry. One is an experiment to determine the distribution coefficient of a metal in a soil solution. Distribution coefficients are chemical parameters commonly used in modelling the fate and transport of pollutants in groundwater systems. In the second experiment, we describe the Soxhlet pollutant extraction procedure, in which commercially available sand provides the needed analyte recoveries and reproducibility. The new experiment described in this teaching exercise describes a very useful experiment for illustrating the fate and transport of a nonretained and retained chemical tracer in a simulated groundwater system. The overall goals of these laboratory exercises are to experimentally illustrate fundamental concepts of contaminant hydrology, specifically step inputs of tracers, pollutant distribution coefficients and retardation factors, and pollutant breakthrough curves.

Rationale and Approach

The lead author of this chapter (Dunnivant) has worked on two major groundwater laboratory and field investigations for the Department of Energy (DOE). When presenting the findings of these investigations at national and international meetings, he was asked why there is not a teaching laboratory exercise for these important fate and transport concepts. The answer is simple-it is very difficult to find a pollutant and a geologic media that utilises these concepts and allows the experiment to fit into the standard three hour teaching laboratory time slot. For example, one investigation with the DOE required several days of continuous monitoring of the effluent from the soil columns, while the second investigation required continuous monitoring periods of weeks to months. Even the most diligent student would balk at these time commitments.

These extremely long experiments present several difficulties in the research lab, most notably designing and maintaining an experimental setup that will operate for the required time without failure; more than one column experiment has prematurely failed. The implications of such failures are even worse considering that the complete effluent profile of pollutant concentration is needed to analyse the data and obtain estimates of pollutant parameters such as the distribution coefficient and soil-specific retardation factor.

The authoritative publication on designing research-grade column experiments is by Relyea (1982) and provides specific parameters, including (1) the required width of the column (based on it being 40 times the average grain size of geologic media), (2) the column length to width requirements to avoid preferential flow along the column walls, (3) water flow rate guidelines, (4) use of an upward flow regime through the soil column, and (5) requirements of the tracer input function and column hardware as to not allow column expansion or cracking of the packed media during the experiment.

In designing a teaching exercise, some of these guidelines had to be modified (such as the pumping equipment for the solutions, column hardware requirements, media grain size, and column expansion or media cracking) in order to minimise the monetary cost and time commitments of the experiment. Modifications needed to obtain our design include slightly reducing the column width requirements (our columns are not 40 times the average sand grain size), the use of a downward flow (necessary for our design), lack of physical packing of the sand media, and a less than conterminous perfect step addition of the tracers to the head of the column (our step input results in a much larger water volume of transition between pure water and tracer than research experiments would tolerate).

The net result of our modifications is a slight broadening of the tracer concentration profile exiting the column, due to increased hydrodynamic dispersion, but the final experimental design is more than adequate to illustrate the most important concepts of pollutant fate and transport in groundwater systems. One final note should be made with respect to the development of this exercise. As you will notice, we use the toxic cadmium ion for our nonconservative (adsorbed) tracer, which generates hazardous waste that must be dealt with at the end of the experiment. Treatment/disposal options are given in the online instructions, but you should consult your college/university safety or hazardous waste officer for further instructions. The use of Cd_{2+} was not by choice but necessity. We attempted the experiment with several other potentially nonconservative metal ions, such as calcium, nickel, zinc, and copper, but these eluted from the column with fluorescein or not at all due to the relatively low solubility product constant (K_{sp}) of the metal hydroxide. Thus, while we tried to design a "green" laboratory, the only metal that consistently worked was cadmium.

Theory and Relationship to Groundwater Transport

Several chemical and physical factors determine the transport of pollutants in real-world groundwater systems. Important physical factors include water velocity and mixing or nydrodynamic dispersion in the system. The most important chemical factor determining the retention of metals in a groundwater system is the affinity of the metal for the soil, quantified by the distribution coefficient, K_a, which is the ratio of the concentrations of pollutant adsorbed onto the soil to pollutant dissolved in the aqueous phase (dissolved phase) or C_{soil}/C_{water}. The adsorption mechanism for metals with soils is usually ion exchange, where naturally abundant metal ions (typically Group I and II, including H^+, Na^+, K^+, Ca^{2+}, and Mg^{2+}), which adsorb onto the negatively-charged sand surface, are exchanged with heavy metals such as Cd^{2+}. This exchange occurs as the front of pollutant passes by a given section of the column. Subsequent flushing with cadmium-free water, containing the Group I and II ions, results in these cations exchanging back for the cadmium. Thus the cadmium, experiencing repeated cycles of adsorption and desorption, proceeds slowly through the column.

Note the relationship between K_d and R in Equation 1 and the relationship or R to down-gradient pollutant concentrations. High K_d values yield high R values, which delay the transport of pollutants, and result in the spreading out of pollutant breakthrough curve. These equations are typically covered in the lecture portion of a contaminant hydrogeology class, and their inclusion in the laboratory setting strengthens the connection between the lab and lecture.

This experiment is divided into two, three hour laboratory periods. At the beginning of the first lab period the instructor introduces the students to the purpose of the experiment, illustrates the various components of the experimental apparatus, notes important points of the procedure, and explains terminology such as pore volume, pulse input, step input, reduced concentration or absorbance, and breakthrough curves. These terms are crucial in understanding pollutant fate and transport in groundwater systems.

Definitions : The pore volume is simply the total volume of water contained between the sand grains in the column. Column-specific pore volumes are measured by the student simply by taking the absolute weight difference of the column apparatus filled with dry sand and the same apparatus filled with water. This value, typically around 30 percent of the total volume of the column, is given in mL for laboratory columns. This term can be expanded for use on the x-axis of a breakthrough curve to essentially give us a time measurement (the more water that has passed through the column at a constant flow rate then the greater the time that has passed.) The use of pore volume measurements gives us a way to compare breakthrough curves from columns of different sizes since each column will be normalised to its own pore volume. The number of pore volumes identified on the x-axis and associated with a given chemical expresses how much water has to be pushed through the column to make each tracer move through the column. One pore volume is necessary to move most conservative tracers through the column, while retained (absorbed) chemicals require addition pore volumes. Again, pore volume measurements on the x-axis are more useful, as opposed to actual mL of water, since each column will have a different volume (mL) of water in the mobile phase, whereas one pore volume always means a unit mobile phase in any given column.

There are typically two simple types of pollutant input scenarios, the step and pulse input functions. A step input is characterised by a change in pollution input from zero to a continuous input concentration (e.g. from 0.0 mg/L to 25 mg Cd^{2+}/L). In this input scenario, the effluent Cd^{2+} concentration begins at zero and increases until it reaches the input concentration (25 mg Cd^{2+}/L in our example). The resulting breakthrough curve will occur later and have a more gradual slope than that for a step input of conservative tracer. Real world examples of a step input would be a leaky underground storage tank or transfer pipe.

In contrast, a pulse input is a one-time, finite input of pollutant (e.g. 0.5 mL of a 25 mg/L Cd^{2+} solution), and elutes as a bell-shaped peak which is shifted later in the experiment relative to the conservative tracer. We originally designed this experiment to include both a step and pulse input of pollutants, but the step input is more easily preformed and the data are easier for the student to interpret. The magnitude of the peak for a realistic pollutant pulse input is extremely small (since the peak is broader due to its longer time in the column and increased dispersion) as compared to the magnitude of the fluorescein tracer peak, and students tend to incorrectly think the Cd^{2+} peak is unimportant. Thus, we have not included the procedure or results for a pulse input experiment.

Plots of the data resulting from a step column experiment consist of the concentration of pollutant exiting the column as a function of number of pore volumes eluted (note in this case the pore volume is actually a measure of time since column volume and porosity are constant, and we are using a constant flow rate). In most cases, we plot reduced concentrations, which express the concentration (or absorbance) of pollutant in a given sample divided by the concentration (or absorbance) of pollutant added to the system. The overall plot is referred to as a breakthrough curve (BTC) since it shows at any given pore volume (or time) the distribution of pollutant in the column, and thus clearly shows "breakthrough" of pollutants from the column.

After the technical terms have been explained to the students, each group of students assembles the sand column and flow apparatus, and takes several physical measurements of the column (weight measurements for pore volume calculations, bulk density, and water content of the column). After testing the setup and upon the approval of the instructor, the student conducts a conservative tracer experiment in which a step input is applied to the column. Samples are collected, and the absorbance of fluorescein (the conservative, or nonreactive and nonabsorbed, tracer) in each sample is measured on a standard visible spectrophotometre. After the complete breakthrough of the conservative tracer, the column is rinsed with distilled water and kept intact (with water covering the sand) for use the following week. During the second week, the student adds a similar step input of cadmium solution to the column. Samples from the column are collected and analysed on a flame atomic adsorption spectrophotometre (FAAS) (ICP may also be used) until complete cadmium breakthrough has occurred (where the effluent Cd^{2+} concentration equals the inlet Cd^{2+} concentration).

The chemistry behind the retention or lack of retention of the tracers is fairly simple. The conservative tracer, fluorescein, does not have functional groups that are readily attracted to the surface of the sand, so fluorescein moves unhindered through the sand matrix. On the other hand, the cadmium ion, has a strong positive charge which is attracted to the negative surface charge of the sand (the surface charge of sand is negative in DI). The cadmium ions are repeatedly exchanged with naturally present Group I ions in the sand, such as Na+, and allow the cadmium ions to slowly migrate through the column.

After the experiment, the students construct the conservative (fluorescein) and nonconservative (Cd^{2+}) breakthrough curves (fractional pore volume of each sample versus analyte concentration or reduced concentration) on the same spreadsheet plot. The retention of cadmium, relative to the conservative tracer, is very evident. Sample data sheets are given in the supplement online instructions. These data are discussed in the following section.

Typical student results for the conservative tracer are plotted. This shows the expected shape of a step breakthrough curve. Note that the conservative tracer elutes from the column first and its transport represents the rate of transport of water through the column. The width or spread of the conservative tracer profile is indicative of the dispersion in the column. Research-grade, well-packed columns will produce very narrow profiles, while more poorly packed columns (such as the ones resulting from our experimental design) will produce broader curves. The Ca^{2+} elution shows considerable delay relative to that of the conservative tracer, and its excess retention time in the column is indicative of the extra time the Cd^{2+} is attached to the sand. In general, if a chemical elutes from a column 2.5 pore volumes after the conservative tracer in a laboratory column experiment, we assume that it will take 2.5 times as long for it to be transported to the monitoring point of interest in a real groundwater system, with comparable mineralogy and chemistry.

Column studies much like the ones described here, but using soil from a specific site rather than pure sand, are used to predict me relative flow or retention of pollutants in real-world groundwater systems. They are also necessary components of laboratory-based remedial investigation-feasibility studies (RI-FS), which are conducted as part of a Superfund site remediation. Once the site-specific velocity of water has been measured or predicted for the groundwater system, the retention results (R and K^d) from the column study can be extrapolated to estimate the transport of pollutants in the real-world system.

Learning Outcomes

We feel this is an experiment of mid-level difficulty that can be used in an advanced undergraduate or graduate laboratory. We have used it successfully in an undergraduate environmental chemistry course and a geochemistry course at a small liberal arts college.

The main outcome from conducting this experiment is that the students gain a practical understanding of pollutant hydrology terminology and laboratory techniques through experiential learning. Specifically, students have a first-hand working knowledge of conservative/nonconservative tracers, the important but complicated concept of pore volume, and are exposed to a very common experimental setup for conducting pollutant fate and transport investigations in groundwater systems.

The most difficult part of the experiment for the student is setting up the apparatus and adjusting the flow and water level in the sand columns, but we feel it is necessary for the student to complete these steps as it enforces the concepts of pore volume and the measurements that go into setting up and constraining such a system. This experience also gives them a small appreciation of the detail required for research-grade experiments. The most difficulty arises during the first lab, when the students are (conveniently) only using the nonretained tracer (conservative fluorescein), and we have found that more than adequate time is available to pack the column at least twice. In fact, the students can probably complete the entire experiment twice during a three hour lab period. By the second week, the students have gained the needed proficiency to repeat the column setup and complete the retained Cd^{2+} tracer experiment.

The calculations required in this experiment are relatively straightforward and most questions from the students concern the initial calculation of pore volume. Thus, we have included a sample data set in the supplemental online file for illustration and practice. The pore volume seems to be the most difficult concept in the lab, and it should be carefully explained in the prelab lecture. Again, pore volumes are normalised quantities representing the amount of water passing through the column, relative to the total amount it can hold. Another was of stating this is: a "pore volume" is the total volume of water residing in the pores of the soil media, while "pore volumes" refers to the total volume of water that has passed through the column divided by the volume of one column pore volume.

This term is valuable because different columns have difference absolute volumes of water, and this allows data (BTCs) from different columns to be overlain and compared when both data sets are presented in pore volumes. Of course we suggest covering the theory of contaminant hydrology in lecture prior to attempting this laboratory exercise. Students always "think" they understand the concepts of pore volume, breakthrough curves, and fate and transport in groundwater systems after lecture, but they gain a much higher level of understanding from these laboratory exercises.

Basic Concepts in Hydrology and Sedimentary Geology

An important goal of education in the geosciences is to instill in students an understanding of process. Geologic phenomena are time-transgressive events that take place in four dimensions, but are typically described using dimensionless devices such as words and mathematics. Students are variously adept at transforming ideas taught through these devices into visualised understandings of the concepts. Knowing the facts, nomenclature, and consensus conclusions surrounding a particular concept does not guarantee real understanding, and real understanding does not require knowledge of academic descriptions. The disconnect between real-world processes and descriptions of these processes is well recognised by geoscience educators, and has prompted the use of model experiments in the classroom that enable direct visualisation of processes in four dimensions. Visual-based learning has been shown to be effective at enhancing the understanding of difficult concepts.

Figure : *Important groundwater concepts include the aquifer characteristics*

Here we describe a scale-model classroom experiment that depicts an earthen dam and reservoir system in cross-section. The basic experiment involves constructing a dam by pouring various sands into the centre of an ant-farm dimension tank, adding water to the 'reservoir' side of the dam, and evaluating the movement and ultimate fate of this water. The experiment clearly illustrates fundamental concepts in sedimentology and hydrology, including superposition, grain-size sorting, angle or repose, hydraulic head, capillary versus saturated flow, plume movement, and catastrophic failure. It is inexpensive to construct (~$50), requires about 30 min. to assemble, and is easy to execute within a single 50 minute classroom period.

The experiment was originally devised for a nonmajor undergraduate course, 'The Water Planet', focusing on the role of water in Earth system processes, but can also be used successfully in K-12 settings, where the focus of the experiment is adjusted according to the level of the students. This experiment is superficially similar to the widely used groundwater flow models (e.g. larger tanks featuring piezometric wells, reservoirs, and aquitards), but differs in that the focus is on both sedimentary processes and groundwater movement, whereas typical groundwater flow models are specialised for teaching more advanced concepts in hydrology. The present experiment is also smaller and more portable, easier to use, and less expensive than the groundwater flow models.

The outline of this chapter is as follows. First, we list the materials necessary for constructing the model and performing the experiment. Next, we outline the basic experimental procedure (variations on this are expected and encouraged) and highlight the geologic concepts that are illustrated at each step. Following this, we suggest possible exercises and calculations to compliment the learning experience, and we conclude with a discussion of concepts learned and broader implications in the context of water issues as a major challenge facing future generations.

'Ant Farm' Tank - The 'Ant Farm' is constructed out of clear sheet acrylic-a material with an ideal combination of rigidity, transparency, and ease of assembly. Dimensions for the six pieces needed to construct a 40 x 15 x 3 cm tank similar to ours are given below in both metric and English units (acrylic thicknesses available in the United States are usually specified in English units):

2 @ 40 x 15 x 0.5 cm (16 x 6 x 0.25 in.)

2 @ 15 x 2.5 x 0.5 cm (6 x 0.75 x 0.25 in.)

1 @ 40 x 3.5 x 0.5 cm (16 x 1.25 x 0.25 in.)

1 @ 40 x 8 x 0.5 cm (16 x 3 x 0.25 in.)

The pieces are welded together using an acrylic solvent and the 'capillary method': two pieces are held together in the proper orientation, and acrylic solvent is introduced along the junction through a hollow needle. Capillary action wicks the solvent between the two pieces, with the solvent dissolving the plastic and evaporating within a few minutes to leave a strong weld. For this method to work, it is important that all pieces are precision-cut to exact specifications and have smooth edges. Most acrylic suppliers will provide custom precut sheets that meet these requirements. Assembly of the tank can be accomplished in less than one-half hour when precut sheets are supplied. This tank is useful for a variety of additional experiments, including demonstration of thermohaline circulation, and as a flat version of the 'Aquifer in a Jug' experiment described by Dowse (2000).

Sand and Other Materials - Sand should be washed and free of clays and silts that impede water flow and cloud the water that pools on the 'downstream' side of the reservoir (unless these phenomena are intended to be illustrated). Fine-grained sand is needed in order to observe capillary water movement, and sand containing a distribution of grain sizes is needed in order to observe grain-size sorting during construction of the dam. We typically use two different sands: a poorly sorted, sub-angular, very fine- to medium-grained quartzofeldspathic sand, and a well-sorted, sub-angular, coarse-grained quartzofeldspathic sand. The former is used for building up the bulk of the dam, and the latter is used as a thin interbed that resists capillary flow, is highly conductive when saturated, and has a steeper angle of repose.

Other materials needed to carry out the experiment are water, food colouring, assorted bottles to contain the various coloured water solutions and waste water, a syringe with a 4+ cm needle for injecting food colouring tracers, a rubber hose and large syringe (~60 cc) for siphoning pooled water from the downstream side of the dam, and a protractor for measuring angle of repose. A funnel is convenient for avoiding sand spillage during the construction of the dam.

The Experiment

Earthen Dam Construction - The dam is constructed by slowly pouring sand into the middle of the tank. We typically begin the dam with the very fine- to medium-grained-sand. The triangular cross-section of the dam develops according to the angle of repose of the sediment being used. Oversteepening frequently occurs and is alleviated by sand avalanches running down the sides of the dam. These events may leave sedimentary traces similar to grain-flow 'sand toes' common in ancient eolian sandstones. Sedimentary sorting occurs as sand moves downward from the apex of the dam, and visualisation of this process is enhanced by using poorly sorted sand. After the dam is about one-third of its final height, we typically switch to coarse sand and construct an interbed that is 1-2 cm thick. This results in a visible increase in the angle of repose, and forms a hydraulic heterogeneity that is important later in the experiment. The dam is finished with the very fine- to medium-grained sand, with the apex of the dam placed about 1/2" below the top of the tank. During the construction of the dam, the concepts of relative age and superposition are illustrated, and it is clear to students why underlying sediments layers are older than overlying sediment layers.

Filling the Reservoir - The reservoir is filled by gently adding water to one side of the tank, taking care to avoid erosive damage to the dam. Food colouring added to the water beforehand enhances the visualisation of water movement, and the adds to the overall aesthetic appeal of the experiment. As water is initially introduced into the system, the water quickly seeps into the pore volume of the dam, and its movement is dominated by unsaturated capillary flow. The capillary flow preferentially follows the fine-grained sand layers with higher capillary potential, including the very thin layers that formed by natural sorting during construction of the dam, and flows uphill along the bedding orientation. At this stage, water movement is severely retarded in the coarse interbeds, where inter-grain pore space provides an effective barrier to capillary wetting.

As more water is added, the dam approaches the point of saturation, and water begins to pool on the 'downstream' side of the dam. At this point, water movement becomes dominated by saturated flow and preferentially follows the coarse sand layers where hydraulic conductivity is greatest. This phenomenon is visualised by injecting concentrated food colouring/water solutions (~ 25:75) into different grain-size layers of the dam using a syringe.

The dye may or may not be immediately visible depending on where the injection is made relative to the narrow dimension of the tank. Dye in the coarse layers moves significantly faster than dye in the fine layers. The phenomenon of 'sapping' frequently occurs at this point in the experiment: the zone in the downstream side of the dam where there is concentrated groundwater discharge becomes eroded by the combination of water flow and high pore-fluid pressure. The amount of sapping increases as the coarse grained, high-conductivity interbed is made thicker.

The injected tracer dyes, if concentrated enough, will have very different density and chemistry compared to the bulk water. This leads to a partial immiscibility between the dyes and bulk water, and the dyes pool coherently on the downstream side of the dam. This phenomenon illustrates the plume behaviour of groundwater contaminants, and approximates the behaviour of the well-studied DNAPL (dense nonaqueous phase liquid) class of groundwater contaminants (Huling and Weaver 1991). Students can attempt to 'remediate' the water by carefully siphoning away the plume using a syringe. This illustrates the fact that, because of dilution, the volume of contaminated water that must be removed or processed is several orders of magnitude larger than the original volume of the contaminant.

Overtopping - The final part of the experiment is the overtopping and catastrophic failure of the dam. Prior to overtopping the dam, the water on the downstream side of the dam should be drained so as to maximise the vertical distance between the top of the dam and base-level. Following this, water should be slowly added to the reservoir side of the dam, such that the actual overtopping begins with the slightest trickle of water over the top of the dam, which inevitably leads to further erosion and a runaway feedback ending in failure of the dam.

Suggested Exercises and Advanced Experiments

The basic experiment outlined above may lead to a variety of questions, calculations, and auxillary experiments geared toward specific groups of students. Obvious calculations are dam volume, bulk porosity of the dam, and water storage in the pore-space of the dam. Measurements of angle of repose can be made during construction of the dam, or a lab period can be devoted to studying angle of repose as a function of grain size and morphology (food items such as rice, beans, etc., are possibilities). Other possibilities include measuring steady-state flow rate through the dam as a function of lithology or reservoir level, or a project focusing on optimal dam design, where each group of students is given identical starting materials (e.g. x amount of clay, y amount of fine sand, a flexible plastic straw, etc.), and the resulting designs are evaluated in terms of lowest conductivity or greatest resistance to over-topping.

Evaluation

Scientific evaluation of the effectiveness of one teaching strategy versus another is an involved process, requiring segregation of demographically indistinguishable experimental and control groups, subjecting each to different teaching strategies, and giving identical tests for evaluating what has been learned. Although such a rigorous evaluation was beyond the scope or intention of the present study, we did perform a qualitative evaluation to gauge students' knowledge before and after the experiment.

Students in the University of Utah's 'The Water Planet' class are primarily nonmajors at the freshman through senior levels, although there are a minority of geoscience majors in the class. Prior to and following the experiment, students are given five graphical questions about topics illustrated by the experiment. The topics include angle of repose, superposition, direction of water flow through groundwater, water-storage capacity of sediment, and recovery potential of water from saturated sediment. The results show that prior to the experiment, students are familiar with the concepts of superposition and angle of repose, do not generally anticipate the possibility that groundwater can move uphill, and overestimate the amount of water that will naturally drain from saturated sediment. Following the experiment, the percentage of correct answers increases for all questions, and most notably students acknowledge the possibility of groundwater moving uphill, and realise that only a very small fraction of water added to dry sediment can be effectively recovered via natural draining of the sediment.

Perhaps more meaningful than the qualitative evaluation outlined above are the results or the end-semester, university-administered course evaluations. Year after year, a majority of students mention the in-class demonstrations when they are asked to describe their favourite aspect of the class. This suggests to us that students enjoy learning through hands-on experiences, and importantly that they remember these experiences. These results are consistent with generally positive comments that we receive from students about the experiments, and the high level of enthusiasm exhibited by the students during the experiments.

In this experiment, students are able to witness first-hand a variety of basic processes in sedimentology and hydrology. Students leave with lasting visual impressions of diverse phenomena such as sedimentary stratification and capillary wetting of dry sediments. The experiment is efficient in the sense that many processes are illustrated within a short period of time. Many of the behaviours of the system are contrary to expectations, including the uphill movement or water, and the coherence of the contaminant plume. Students are more apt to believe that these phenomena really do occur if they are able to see them happen for themselves.

A major goal of geoscience education is to inform and educate the public about socially and environmentally relevant issues. This experiment naturally brings up a discussion of water issues. What is the ultimate fate of groundwater contaminants? What happens to water that enters a reservoir? How much water might be lost to seepage into bedrock? Under what circumstances does water "flow uphill"? Water availability is a major issue confronting humanity, and it is important for students to think about water in the context of real geologic processes. Students in geoscience classes are voters and future policy makers, and learning experiences such as the one described here are a contribution towards wise decision-making.

Conceptions of Scale Regarding Groundwater

The major science frameworks and standards documents used in the United States explicitly and disproportionately focus on surface oriented hydrologie processes. These documents do, however, contain a relatively implicit and subtle call for appropriate student understanding of groundwater principles by the repeated and explicit inclusion of the water cycle in content standards across elementary, middle, and high school settings. Consequently, some states' standards documents include hydrogeologic concepts that should result in the inclusion of those concepts in K-12 science instruction. Additionally, hydrogeologic principles are routinely incorporated in post-secondary introductory geology coursework and serve as one of the fundamental components of hydrogeology courses.

Figure : *Water: large-scale desalination. Having realised that its groundwater*

At every one of these points of instruction throughout a student s formal education, it is the role of the teacher to assist the student in constructing conceptions that are perpetually more consistent with those held by the scientific community. Altering students' ideas, however, about scientific concepts is complex and based on a wide range of factors. In response, educational researchers have developed generalised models of how this change may occur. Posner, Strike, Hewson, and Gertzog (1982) asserted that for conceptual change to occur, four conditions must be met: 1) students must be dissatisfied with their current idea, 2) the new idea must be understandable, 3) the new idea must be reasonable, and 4) the new idea must be fruitful in new contexts. This basic model is widely accepted in the science education community and many researchers have proposed slight variations.

Strike and Posner (1992) have emphasised that this model is not intended to be directly applied as a "prescription for instruction", but rather generalised elements necessary in changing people's conceptions. Science educators have incorporated or identified some or all of these elements in many different constructivist-based instructional strategies including learning cycle approaches. Usually one of the first steps in conducting a learning cycle involves assessing students' preconceptions. These ideas are almost always inconsistent with those of the scientific community and are consequently termed alternative conceptions.

Identifying preconceptions and/or alternative conceptions is important in altering students' ideas because it informs the decisions teachers make regarding how they choose to address the elements involved in conceptual change. For example, due to the well-documented tenacity of alternative conceptions and the numerous sources of origin of those conceptions including but not limited to analogy, ontology, instructional materials, and teacher competence, informed educators will appropriately vary the duration of the lesson, instructional tools used, and/or instructional strategies implemented to effectively teach a given concept.

In the case of groundwater, like many geoscience concepts that are often directly unobservable, teachers will often employ the use of field trips, hands-on experiences, experiments, physical three-dimensional models, mathematical formulas, static two-dimensional images, and animations in addition to text to produce meaningful learning in their classrooms and address students' preconceptions. These visually-based instructional methods and tools are employed in an effort to assist students in developing an appropriate mental picture of groundwater principles. Such mental pictures constructed of a variety of conceptions are typically described as mental models as opposed to conceptual models that are typically mathematical in nature and are "precise and complete representations that are coherent with scientifically accepted knowledge".

For the purposes of this study, we are operating from a generalised definition of mental model, which we consider to be a particular set of relationships between conceptions. To use an analogy, we see the mental model as a house and the individual conceptions as the studs, nails, plywood, shingles, etc. In order for the mental model (house) to be consistent with the scientific model (a house that looks exactly like a real house), the various conceptions (plywood, nails, etc) must also be consistent with those held by the scientific community (building supplies that look exactly like real building supplies). For instance, if a student thinks that all rock is completely solid and there is no such thing as tiny spaces within it (considers plywood made out of light-weight foam), then the mental model constructed of groundwater (house built with light-weight foam plywood) will not be consistent with (look or function like) the model held by the scientific community (house built with wood plywood). Even if the conception is appropriate (plywood is made out of wood), if the scale is inappropriate (the plywood is 1000 metres thick and a kilometre in length), the resulting mental model (house) will also be inappropriate (look unlike a real house). As such, mental models inherently possess an impediment to the progression towards a complete and appropriate scientific model because of the essential inclusion of the absolute and relative scales of the conceptions comprising the model. The effective teaching and learning of scale has historically been considered problematic, but is extremely important in a variety of geoscience contexts. The spatial relationships between groundwater concepts like porosity, permeability, aquifers, and flow regimes are crucial to the development of appropriate mental models, because inappropriate scale applied to any one of the associated conceptions results in an inappropriate model. As such, we examined how different student populations apply scale to their conceptions of groundwater principles.

This chapter reports on one aspect of a larger research study conducted to examine understandings of groundwater principles and processes held by children and adults. The study is limited in scope (i.e. small sample sizes from one location) and all assertions are made solely within the context of the sample. The insights gained, however, will aid in developing and refining instruments designed to assess groundwater-related understandings and will serve to inform instructional practice. Larger scale studies are currently in progress to gain a more complete and accurate picture of the teaching and learning of groundwater both in the United States and internationally.

The study population consisted of the purposeful sampling of secondary and post-secondary students. Our intent was to select populations with varying levels of interest in and understanding of groundwater. The secondary participants consisted of twenty-nine twelfth-grade Anatomy and Physiology students in a large public high school located in a major coastal city in the southeastern United States. Their teacher indicated that the majority of the students anticipated pursuing post-secondary studies. The students reflected the general school population in all other ways with the exception of gender because twenty-three of the twenty-nine participants were female. Ten of the participants had successfully completed an earth/environmental science course during their secondary school experience and two reported that they had taken a course that focused on groundwater.

The post-secondary participants consisted of two groups of students enrolled in interdisciplinary environmental studies courses at a liberal arts university located in a major coastal city in the southeastern United States. Group 1 consisted of thirty-two students, that included three freshmen, five sophomores, eleven juniors, twelve seniors, and one graduate student all enrolled in one of two elective environmental studies courses that had no geology prerequisites. Sixteen different majors, including 19% of the participants reporting an earth/environmental science related major (n=32), and seven minors, including 50% of the participants reporting an earth/environmental science related minor (n=32), were represented among the students. Two students reported having engaged in scientific research regarding hydrology and two reported that they had taken a course that focuses on hydrology in their post-secondary education.

Group 2 consisted of twelve students, that included two sophomores, four juniors, one senior, and five graduate students all enrolled in one of two introductory hydrogeology courses. The educational backgrounds of the graduate students included holding a B.S. degree in Geology (n = 2), a B.S. in Physics (n = 1), a B.A. in Anthropology (n = 1), and a B.S. in Biology (n = 1). Four of the graduate students reported having participated in hydrogeology research and one reported completing a hydrology course in the past (i.e. Wetland Hydrology).

The teacher from each class administered the instrument to their students during their normal class time. We collected data from the secondary students the last week of classes. The post-secondary students completed the instrument during the first week of the course. All participants completed the instrument entitled Groundwater Survey. The indigenous instrument is composed of a background information section including questions regarding gender, year in school, and prior earth/environmental science and groundwater coursework. The remainder of the instrument is composed of multiple-choice items that deal with issues of structure, scale, and perceived importance of groundwater. In addition to the instrument answer choices on the first and second items, participants had the opportunity to provide an answer or their own using a fill-in-the-blank response option. The development of the first instrument item was informed by earlier studies that identified a variety of descriptors used by the general public when discussing groundwater. When we developed the distracters for Item #1, we selected from among these common descriptors (i.e. pools, lakes, rivers, streams, etc) and included the one considered most used by the scientific community (i.e. pores/cracks). In order to collect data regarding how students applied scale to these specific descriptors, we developed Items #2 and #3. Participants had the option of selecting more than one descriptor for Item #1, so we provided scale choices for each of the possible descriptors in Item #2. The development of the options was based on previous scale research that suggests students may use familiar objects when describing the relative scale of natural phenomena and from data collected during interviews in previous groundwater conceptions studies. The choices available in Item #3 were constructed to provide researchers with data regarding students' notions of vertical scale. Researchers chose the depth ranges to discriminate between students who held reasonable notions of vertical scale and those who did not. Items #4, #5, and #6 consist of parallel questions designed to illuminate any disparities in perceived importance of groundwater among teachers, voters, and themselves.

Two faculty members from the geology department at a university in the southeastern United States who specialise in hydrogeology completed the Groundwater Survey instrument and provided responses that established face validity. Researchers also collected data to support the validity of the instrument from the students' perspective. In another phase of this research project, data were collected and triangulated using a variety of methods and instruments that included a modified version of a previously published groundwater conceptions instrument (i.e. a drawing prompt), the instrument included in this study, and transcriptions of videotaped think aloud interviews. The think aloud data were considered in this study as a means of instrument validation to ensure that the research team accurately interpreted the words used by the participants. Content validity was established within the context of the participant group as evidenced by interview excerpts explicitly addressing notions of scale (i.e. from the think aloud interviews) and numerical annotations contained in the drawing prompts (i.e. the previously published instrument), all of which were consistent with the participants' responses on the instrument employed in this study. Because all participants were not interviewed, however, due to resource limitations (e.g. time) and due to the small sample size, we make no claims regarding the instrument's validity in other contexts. It should be noted that while this study represents an initial attempt at the development of an instrument designed to assess student understanding of scale in the context of groundwater, additional large-scale studies and instrument refinement are necessary to truly develop a valid and reliable instrument for use with diverse populations. The quantitative methods we employed in this study incorporated simple descriptive statistics (i.e. frequency counts and percentages) of item responses.

The results provided an indication of participants' conceptions regarding the storage of groundwater, the scale associated with a given storage feature, vertical scale of wells, and perceived importance of knowing about groundwater. The responses we gathered from the establishment of face validity are considered to be consistent with those held by the scientific community. As such, most groundwater in the eastern part of the United States is considered to be stored beneath the ground in pores and cracks, ranging in size from microscopic to the size of an eraser on a pencil. Additionally, most human drinking-water wens are considered to be shallower than 5,000 feet.

Perceived Importance of Knowing about Groundwater - It was necessary to include the items (i.e. #4, #5, and #6) that addressed the importance of knowing about groundwater because they directly assessed the perceived value of the topic and indirectly assessed participant interest. Both the perceived value of and interest in a topic are influential factors that affect conceptual change, because they directly relate to whether a person considers the concept to be fruitful.

The secondary Anatomy and Physiology participants provided responses that indicate a much lower perceived importance of knowing about groundwater than either of the post-secondary hydrology-oriented participants. The post-secondary participant groups' responses were very similar to one another and both groups consistently assigned a high degree of importance to all item groups (i.e. themselves, teachers, and voters). Interestingly, the secondary students ascribed a greater degree of importance of knowing about groundwater to teachers and voters as compared to themselves. We analysed the secondary participants' responses further because some participants and completed an earth/environmental science course.

Approximately a third of the secondary participants reported that they had taken an earth/environmental science course in their secondary school experience. This subset of the secondary participants provided very similar responses to those who had not taken an earth/ environmental science course for Items #1, #2, and #3. The exceptions were the items regarding the participants' perceived importance of knowing about groundwater. In general, participants that completed an earth/environmental science course did assign a greater degree of importance of knowing about groundwater for themselves than did those who had not taken an earth/environmental science course. The trend, however, of ascribing a lesser degree of importance for themselves as compared to teachers or voters existed in both subsets of the secondary participants.

Conceptions of Scale - Participants' conceptions of vertical scale were directly assessed through Item #3. The Group 2 participants (100%, n=12) and the majority of the Group participants (88%, n=32) indicated that the depth of most human drinking-water wells in the United States are less than 5,000 feet deep (i.e. the first two distracters in Item #3). In contrast, thirty-four percent of the secondary participants (n=29) thought wells were deeper than 10,000 feet.

In general, participants described the ways that most groundwater is stored underground using a variety of terms. These results are consistent with other studies regarding language commonly used to describe groundwater. Furthermore, there is little difference among the percentages of responses selected between participant groups despite the substantial differences in formal geoscience education and levels of interest in and value of groundwater. The largest differences occurred with the terms "pipes" and "pores/cracks". Almost a quarter (23%, n=65 responses) of the responses from the secondary participants were "pipes", as opposed to eight percent and sixteen percent from Group 1 and Group 2 respectively. Group 2 had the highest percentage of responses (34%, n=32 responses) for "pores/cracks", in contrast to Group 1 (32%, n=84 responses) and the secondary participants (20%, n=65 responses).

Three participants out of seventy-three total, two from Group 2 and one from Group 1, provided responses that matched those deemed consistent with the scientific community. Two of the three participants were graduate students and one was a sophomore. One of the three also reported actively engaging in hydrology research. All the other participants (n=70) provided either, multiple responses, one inappropriate response regarding storage of groundwater, or an inappropriate size range for pores and cracks. The variability in size ranges for a particular structure, even within participant group is considerable. For example, most participants who provided a response of "pools/lakes" described them as being "house - skyscraper" large. Interestingly however, twenty-eight percent of the responses from 12th graders, forty percent of the responses from Group 1, and thirty-three percent of the responses from Group 2, indicated that "pools/ lakes" were "basketball/beach ball car" size or smaller. The majority of the participants that selected "streams/rivers" as a response again described the structure as "house - skyscraper" large, while thirty-five percent of the responses from 12th graders, 1 forty-eight percent of the responses from Group 1, and twenty-five percent of the responses from Group 2, indicated that "streams/rivers" were "basketball/beach ball - car" size or smaller.

The descriptions of size regarding the response "pipes" were more evenly distributed than the descriptions of size of the previous two structures, and included forty-two percent of the responses from Group 1 and forty percent of the responses from Group 2 selecting the size range "microscopic - eraser on a pencil". While most participants who provided a response of "pores/cracks" described them as being the size of "microscopic - eraser on a pencil", forty percent of the responses from the 12th graders, thirty-rive percent of the responses from Group 1, and thirty-six percent of the responses from Group 2 provided a description of size that was larger. Some participants also provided responses to the fill-m-the-blank option of "other". Three secondary participants included other structures such as "reservoirs" of every size option available, "wells" the size range of a house to a skyscraper, and "craters" the sizes of microscopic to an eraser on a pencil and the sizes of a house to a skyscraper. Four Group 1 participants included "reservoirs" the size range of a house to a skyscraper and the term "aquifers". The three participants who wrote "aquifer" each described the size differently such that all three size ranges provided on the instrument were used. Two participants from Group 2 also provided "aquifer" as a response. One described an aquifer's size as other: large/ sq miles", while the other participant described the size range as "basketball/beach ball-car".

The results of this study indicate that the participants held inappropriate conceptions of hydrogeologic principles despite groundwater's importance to their health and economic well-being. They describe groundwater storage using multiple structures other than pores and cracks. Although Dickerson and Dawkins (2004) documented that the use of such language (e.g. underground pools and rivers) can represent both appropriate and inappropriate conceptions, the application of scale helps further articulate those conceptions. Participant responses regarding the size ranges of groundwater storage structures show that the participants possess a wide range of ideas concerning scale. Many selected sizes for the groundwater structures they chose that mirrored the surface analogues. For example, participants generally described underground pools and lakes and underground streams and rivers as large features just as they appear on the earth's surface. Likewise, underground pipes were attributed sizes comparable to those of pipes we see in houses and at construction sites. Pores and cracks were also considered small in size by most of the participants that selected that structure. Equally important, however, are the number of participants who chose sizes that deviate from the scale of the surface oriented analogue. For example, approximately a third or more of all those that selected underground pools and lakes and underground streams and rivers indicated that those structures were smaller than a car and some even considered them smaller than an eraser on a pencil. Interestingly forty-two percent (n=12) and forty percent (n=5) of the responses regarding underground pipes from Group 1 and Group 2 respectively labelled that structure as small as microscopic and as large as the size of an eraser on a pencil. Lastly, approximately a third of all those who selected pores and cracks chose sizes for that structure larger than a basketball or beach ball. Several students even applied scales on the order of houses and skyscrapers to typical pore and crack structures. It is likely that mental models erected according to such alternative scale conceptions are inappropriate.

Considering these findings it may be more important, in terms of mental model development, for the scale to be consistent with the actual in-situ environment than for the language to be consistent with that of the scientific community. For example, a student's conception of underground pools and lakes that are microscopic to eraser sized may not adversely affect the construction of an appropriate mental model if those terms imply all of the parameters of the term pore space.

Conversely, if students use scientifically acceptable language like pore space but think that most groundwater exists in pores that are a half a kilometre in diametre, when they combine that inappropriately scaled conception with their other conceptions of permeability, aquifer, etc, the result is likely to be an inappropriately constructed mental model. In addition, a disconnect may exist between groundwater principles in some of the participants' minds that may adversely affect mental model development. For example, some participants selected the option "pores/cracks", but also described the storage structures of most groundwater in the eastern United States as "aquifers" and "wells" by selecting the distracter of "other" and providing a fill-in-the-blank response in Item #1. This raises the question of whether the students understand the relationship between the two, however further research is necessary to determine if this is the case.

Since students may enter courses holding ideas of groundwater that are inconsistent with those of the scientific community, initial assessment is prudent. Among the more surprising findings in this study is the number of undergraduate and graduate geology majors that enter hydrogeology courses with notions of groundwater storage and scale that dramatically diverge from those held by the scientific community as illustrated above. These findings emphasize the need for early assessment of students' conceptions in order to know where to begin instruction. Assessing students' conceptions and the mental models they construct from them, however, is a difficult task. It is important to begin the development or selection of assessments with the understanding that the development of appropriate mental models involves issues of scale. While vocabulary can be conceptualised a number of ways, sizes based on familiar objects provides insight into how the vocabulary is being used and the appropriateness of the concept. Such assessments are also important in exacting conceptual change and facilitating the development of appropriate mental models of groundwater by making students cognizant of the disparity between their mental models and the scientific community's model. For example, requiring students to articulate their initial mental models in pictures and words provides the instructor with insight into how much or little their students already know while providing a point of comparison regarding the scientific community s model. The role of assessment grows in import based on our results that the resiliency of the participants' alternative conceptions of groundwater scale was strong even after completion of introductory undergraduate geology courses. This is particularly troubling because very few people, including K-12 teachers, pursue advanced coursework in hydrogeology, yet an understanding of potable water (i.e. where it comes from and how we get it) is certainly part of basic scientific literacy, and scientific literacy is something every vote-casting individual should possess.

Instruction should also incorporate a variety of strategies that promote personal relevance construction. Interestingly, a trend emerged among the secondary students in which they ascribed a higher degree of importance of knowing about groundwater to teachers and voters as compared to themselves. These findings possibly indicate that these participants do not view themselves in the roles of either teachers or voters. The results are not completely surprising considering the age of the participants, yet are important to consider when addressing socio-scientific issues in formal education contexts for the purposes of promoting conceptual change. For example, many science teachers use debates and role-playing activities to explore issues of groundwater usage and pollution and to teach groundwater principles. Addressing socio-scientific issues in the classroom may be an effective way to get students engaged, build personal relevance, and facilitate conceptual change, but only if students view the roles as viable. So promoting the use (e.g. voting) of their newly constructed understandings in a personally and socially responsible way is important.

As evidenced by the variety of participant selection of groundwater storage structures and sizes, the use of abstractions (e.g. language) complicates teaching and learning with respect to groundwater. As such, strategies and tools that emphasize visual information and deemphasize vocabulary may be more successful. Examples of such strategies would include common visual experiences such as field trips, teacher use of drawing, use of core samples, manipulation and development of physical and three-dimensional computer models in cooperative learning groups. All of these strategies should also incorporate the use of student assigned descriptors instead of scientific language and explicit instruction regarding scale through direct measurement by students and/or the teacher, visual cues such as labeling, and group discussions and lecture regarding the common visual experiences. After students have begun developing appropriate groundwater conceptions through the use of common visual experiences and explicit instruction regarding scale, scientific labels can then be added.

Further research is needed regarding students' conceptions of groundwater and the ways that they integrate those conceptions to build mental models of subsurface hydrology. Based on the results from this study, which suggests that participants' applications of scale to groundwater concepts were often inappropriate, researchers may consider increased attention to scale components during future conceptions studies. Further refinement of instruments designed to quickly and accurately assess students' notions of scale are also needed. The teaching of hydrogeology concepts is difficult. Abstraction and perceived irrelevance serve as impediments to conceptual change concerning groundwater. This difficult situation requires that teachers attend to a wide variety of students' conceptions and the relationships between them, paying particular attention to issues of scale. It is also important that teachers provide a context in which the mental models constructed from such conceptions can be applied to individually and socially relevant questions and issues from the students' perspectives. By attending to both scale and relevance in the context of groundwater teaching and learning, we can illuminate the mystery of hydrogeology for all students.

Learning in Geo-hydrology

Changing preconceptions is among the most challenging tasks that today's science educators are dealing with. It is not only a matter of making updated materials available to students, but also is an issue of presentation style and its mechanisms. The literature of science education has been flooded with innovative techniques of delivering lectures in a classroom and involving students in interactive discussions. It is only in the early 80's when hands-on activities emerged as a precondition of improved learning of undergraduate students. Nevertheless, in the recent science-education reform efforts, teachers are constantly looking for innovative ways of involving students in practical exercises in the laboratory.

The objectives are two-fold: show them how science works, and involve them in meaningful activities that can change their misconceptions about science. As a result, numerous lab activity books have become commercially available in the last two decades, which include hundreds of pages of illustrations and work schedule. But, some of these lab activity books are so complex that the students get bored and become more scared of science. Indeed, field-based training of undergraduate students in environmental sciences is an area that needs to be strengthened, because having a field perspective is the only way to understand and apply environmental and hydrogeologic principles. Many environmental programs neglect field-based training in favour of computer-modelling exercises, which are abundant and commonly the only experimental hydrogeological or environmental activities offered to students.

An outdoor experimental setup (on-campus or near campus) can provide students a smooth transition from classroom learning to the real world environment. Rahn and Davis (1996) used an operational well field consisting of a main well and 14 observation wells as an educational and research facility for students at South Dakota School of Mines and Technology. They found that students had a better understanding of well hydraulics and showed much enthusiasm during these outdoor educational activities. They learned about static water level, potentiometric surface, pumping, slug and tracer tests, drawdown etc. Trop et al. (2000) successfully integrated field observations with laboratory modelling for understanding hydrologie processes in an introductory Earth-Science course at Purdue University. They took the undergraduate students to a campus water-well field located less than a mile from the classroom. The instructors first showed them the water wells that supply most of the drinking water for the university community and then generated some discussions regarding the source of the water and the physical characteristics of the aquifer. The students were then taken to a sand and gravel quarry to study Pleistocene deposits that form the uppermost part of the above aquifer.

It was clear that the opportunity to visit the water-well field and the quarry removed some of their misconceptions. The students were then ready to go back to the laboratory and make their own aquifer models. Similarly, Noll (2003) showed the effectiveness of field observations in understanding hydrogeological parameters at State University of New York College at Brockport. Students were taken to a sand quarry where they observed in three dimensions the natural variations in the subsurface that would otherwise be hidden if they were to use only core samples from an unconfined aquifer. The students looked around, discussed their observation, and then collected soil samples for analysis. In the laboratory, they analysed these samples for moisture content, bulk density, particle density, and particle size distribution. In addition, the students determined hydraulic conductivity in the field using percolation tests. The instructors found a positive impact of these activities on student learning. Numerous other investigators showed that field experiences are an important component for understanding hydrogeological concepts. This demonstrates the effective use of outdoor teaching laboratory, namely two water monitoring well sites, to enhance scientific literacy in undergraduate students. The main component of the project was conducted at the University of Northern Iowa campus in Cedar Falls and was funded by the National Science Foundation and the Iowa Science Foundation. Additional activities were done at the State University of New York at New Paltz for comparison. The expenses at New Paltz were covered from their internal faculty research support program.

Project Development

Pedagogical Approach - Experiential learning should be an important component of undergraduate education. For instructors, factual examples facilitate effective teaching. It is not only important what students should learn, it is equally important how science concepts are taught. Appropriate materials should be used to present a scientific concept. In order to initiate their thought process, students must see the tools that are integral parts of the concept. Appropriate teaching mechanisms can indeed change preconceptions. When learning process begins, the students naturally apply their own critical thinking set in their mind and compare the new concept to the internal structures (preconceptions) that are already present.

If the new experiences don't fit their original concept, the curriculum must be flexible enough to reconsider what had been assumed and to rebuild their internal structures. Donovan and Bransford (2004) highlighted in their book "How Students Learn" the essential role of factual knowledge and conceptual frameworks in understanding complex systems, and the fact that learning is constructed on the foundation of existing knowledge and experience. Investigators reported considerable success in educating students where teaching and learning methods were used in an interactive environment. Such an environment of teaching and learning, which is widely known as constructivism, may not De as traditional as lecture-oriented education program, but it presents strong experimental data for effective science education. Constructivism provides students with direct sensory experiences and encourages them to question their preconceptions. Subsequently, students learn how to develop hypotheses, interpret data, offer constructive arguments, and make predictions. Above all, it promotes the basic elements of science standards among students, which includes curiosity, observation, synthesis of observed data, reasoning, and objective conclusions. The whole process also allows students an opportunity of teamwork, which is a very important component of effective learning. While working on their assignments, they learn from one another to fill the gaps in their understanding of science.

Based on the important elements of science education as discussed above, the following pedagogical approach was taken in all interactive learning sessions in this project:

Scientific Inquiry - The students were asked interesting questions and encouraged to respond on the basis of their collection of evidence. They were asked to work in groups, share ideas, and de-emphasize memorisation.

Scientific values - The students were told that curiosity plays an important role in learning science. In the process, they were rewarded for their creativity and flexibility in ideas.

Learning anxieties - The students were encouraged to build on their success. They were given proper training in using lab equipment and were told that instruments could help them find a precise solution to scientific problems.

Dissemination - The students were encouraged to continue to explore and excel in what they do. Also, they were asked to let others know what they have learned.

Teaching Methods

University of Northern Iowa - We applied the teaching methods to the following three courses at the University of Northern Iowa; (i) Environmental Hydrology (15 students; juniors/seniors): The primary focus is on surface water pollution and watershed nydrology; (ii) Hydrogeology (12 students; juniors/seniors): The class teaches principles of hydrogeology, with a major focus on the principles of chemical Hydrogeology; (iiij Physical Geology (General Education course; 75

students; primarily freshmen/sophomore): The primary focus is understanding the earth s dynamic systems and its physical environments. In order to develop hydrologie concepts, a 3-step approach was taken. The methods are briefly described below.

Step 1: class lectures - The students were exposed to hydrologie concepts through detailed classroom lectures. As a part of in-class assignments, they were given written exercises dealing with Darcy flow concepts. On given diagrams, they calculated stream discharge, measured porosity/permeability, estimated groundwater resources, and studied the mechanisms of contaminant transport. They constructed flow nets and answered questions on water quality issues. All classroom activities were driven toward understanding the following concepts: (a) Groundwater is contained in pore spaces and fractures; (b) Groundwater flows from high hydraulic head to low hydraulic head; (c) Aquifers are recharged by precipitation; (d) Human activities can contaminate groundwater. (e) Groundwater quality can vary temporally and spatially; (f) Wells can serve as point sources of groundwater pollution; (g) Contaminated groundwater can pollute surface water and vice versa.

Step 2: laboratory simulation - This component involved lab activities with a groundwater flow simulation model. These hands-on activities were targeted toward validating the concepts learned so far in the classroom.

Students performed all simulation activities under a set of guidelines from the instructor. The students were divided into groups to appreciate the importance of teamwork in science and develop self-confidence. The process of concept development was as follows:

(a) Concept: Groundwater wells can serve as point sources of pollution.

(b) Student activity: The students injected dye tracers as well as bromide and chloride-tagged waters through injection wells in the simulation system and then discharged water through pumping wells. They simulated water sampling from wells, a leaky lagoon, and a stream that were facilitated in the model. Subsequently, they chemically analysed the water to understand the hydrologie link among groundwater wells, rivers, and land sources of pollution. They injected coloured dyes in both the confined and the unconfined aquifers in the model, monitored travel paths, and estimated groundwater velocities. Additionally, they measured hydraulic gradient of the water table in the simulation system.

(c) Follow-up discussion: Wells with cracks or rusted casings, or wells not properly sealed at the surface, can serve as conduits for farm chemicals as well as contaminated surface water to enter the aquifer and cause groundwater pollution. Notice that one of the pumping wells is contaminated with chloride only and the other is contaminated with only bromide. This is because the pumping wells are completed in two different aquifers. The clay-confining layer restricts flow between the two aquifers. Wells should be protected from damage, and should be properly sealed when they are to be permanently abandoned. Refer to state and county regulations for codes of construction, maintenance, and abandonment of wells.

Step 3: outdoor experimental site - The students were subsequently given access to an outdoor experimental plot in order to enhance their understanding of geohydrologic concepts. Here the students had direct access to groundwater monitoring wells and a nearby stream (called the "Dry Run Creek). They solved varied hydrologie problems, which were based on their new concepts developed through the previous two steps.

The well cluster at UNI has 8 shallow wells (3.7 to 6.1 m deep) and 2 deep wells (22.9 m and 27.4 m). Students walked to the site during lab sessions and performed ground water as well as surface water monitoring activities. All activities were preceded by onsite discussions of well depths, types, and the drilling method. The shallow wells are constructed with PVC tubes in the alluvial aquifers above the pre-illinoian till. The deep wells are metal cased and are completed in the upper Devonian carbonate bedrock aquifers. The bedrocks in this area directly underlie the pre-illinoian till. Before installing the well cluster, a formal application was filed to the Physical Facilities Planning Department at UNI to acquire this 7.6 m x 7.6 m land. The piece of land was granted in consideration of the long-term educational goals of the university. Following this, a curriculum development proposal was submitted to the Iowa Science Foundation requesting funds for the on-campus well cluster. Subsequently, the project was funded. The drilling job was then contracted out to a state-licensed driller upon approval by the university. Interested students went to me site to watch the rotary drilling procedure, which took 2 days to complete. Later, the National Science Foundation (NSF) granted additional money to conduct student activities on this site. The Dry Run Creek flows within 9 m of the well cluster, which was indeed a major factor in selecting the well site location. It gives the students an excellent opportunity to study both ground water and surface water.

The students formed teams on site and collected water samples from the wells and the adjacent creek. For the lower division undergraduate students enrolled in the Physical Geology course, the workload was less rigorous. They sampled water from the wells as well as the creek and analysed them for pH, TDS, Conductivity, Dissolved Oxygen, and temperature. In addition, they used test strips onsite to measure chloride, nitrate, and sulfate dissolved in water. In fact, they spent much of their time in discussing their findings. The instructor was directly involved with them in dialogues as they were collecting data. The discussion sessions primarily addressed the possible sources of the dissolved materials in water. Before going back to the lab, the instructor made sure that the students understood why the chemistry of the ground water and the stream water were different. For the upper division students enrolled in the Environmental Hydrology and the Hydrogeology classes, the workload was more rigorous. In addition to doing all the above, they used a depth indicator to measure the depth to water table and then developed a flow net in the vicinity of the well site. Then based on the flow net, they answered questions regarding well contamination. For chloride, nitrate, and sulfate, instead of using test strips the students took the samples back to the lab and analysed them by ion chromatography.

Instructor discussed with them the method of ion exchange as individual peaks appeared on the computer monitor. This was a much-needed activity for this group because most employers in the field of hydrology want applicants to be familiar with common water analytical equipment. In another day, this group of upper division students went back to the site and repeated the activities by visiting 4 other spots along the Dry Run Creek. They used their analytical kits to study water chemistry and understand how surrounding land use can contaminate stream water. They calculated stream discharge at 2 spots by measuring channel cross sections and water velocities. All activities were followed by discussions. They soon began to understand the cause and effect relationships in the hydrologie environment.

State University of New York, New Paltz - At State University of New York (SUNY), geo-hydrologic activities that were comparable to those conducted at UNI were introduced. The primary reason for replicating these activities at SUNY was to verify the results across academic institutions.

Exercises were developed to integrate the theory, field data collection, and data processing. For this purpose, the hydrologie field site at the Institute of Ecosystem studies in Millbrook, New York, was used. In addition to the basic "hands-on" activities, the students conducted pump test at this site. The pump test data were then used to develop a ground water flow model of the monitoring wells network site. In order to develop the model, the students used the real-world hydraulic properties of the aquifer derived from the pump test data. A 72.5 m deep ground water well was later installed on SUNY New Paltz campus in October 2004 by the US Geological Survey of Troy (USGS), New York. This well was drilled by the USGS as their ground water monitoring well network under the educational outreach program. Interested faculty at other campuses could contact Local USGS offices to explore the possibility to have free installation of on-campus ground water monitoring wells. The system of hydrologic monitoring network at SUNY, New Paltz involved several groundwater wells, a rain gauge, current flow metres for stream gauging, and tensiometres to monitor the unsaturated zone.

Unlike using a textbook or preexisting data, students used their own real world data collected in the field. They gained first-hand experiences in drilling and installing groundwater wells, designing the monitoring system as well as collecting, describing, and testing subsurface materials. They followed up first by analysing the data they had collected, and then by modelling the actual groundwater flow. This approach helped the students understand the difficulties in building a model from scratch as opposed to textbook simulations where running the model never leads to unexpected outcomes. Students also integrated the principles and practices of surface water studies, such as sampling and monitoring, quality control and assurance, and use of water-quality equipment. They learned how to calibrate and use electrochemical probes to measure pH, specific conductance, temperature, and dissolved oxygen. The activities dealing with the hydrologie monitoring system were integrated into the courses as follows:

(a) Physical Geology (150 students): This is a general education course. Science and nonscience majors were introduced to basic ground water concepts through discussions onsite. Subsequently, they measured water levels at the on-campus well site.

(b) Hydrogeology (20 students): Most or these students are geology majors. They were involved in detailed "hands-on activities, which included determining groundwater flow direction, water quality sampling, pump tests, slug tests, and tracer tests. Much of their classroom learning was clarified through these activities.

(c) Advanced Hydrogeology (15 students): Students in this course are all geology majors who had taken most of the basic geology classes. They worked on a field based hydrogeologic site characterisation exercise in this project. The exercise included determining groundwater flow direction, water quality sampling, pump tests, slug tests, and tracer tests. Students also developed a groundwater flow model of the study site using their own data collected in the field.

(d) Field Hydrogeology course (20 students): This is a 1-credit course offered in the Spring Semester every year. The course is open to students from other campuses. The field activities are usually conducted for two days over the first weekend in May. This course provides an excellent opportunity to disseminate the results from SUNY to other campuses. Students worked on hands-on exercises relating to drilling and installing groundwater wells, and then designing a groundwater monitoring system. The primary focus of this field course was to have students work in small groups of 2 or 3 in selecting monitoring well locations and installing 3 or 4 piezometres. The activities provided them with a better understanding of the shallow groundwater flow directions. The piezometres are inexpensive, which were inserted into the ground by using a large hammer.

Assessment of Student Learning

Evidence of improved critical thinking skills and reasoning of students were demonstrated on essay questions, and critical problem solving questions. Field-oriented realistic questions were incorporated in the subsequent exercises and their responses were evaluated. Science educators have long been in agreement that a major goal of science education ought to be fostering skills of scientific thinking. The students must be able to define a hypothesis, test hypothesis, and evaluate evidence. The impact of the 3-step teaching approach in changing misconceptions among students in this project was evaluated, the details or which are discussed below. It was observed that in the process of discarding preconceptions, students acquired new information and then reorganised them in their existing knowledge base. In addressing this issue, the analogy between individual learning and conceptual change in scientific disciplines has always been useful in developing a suitable framework for analysing science learning.

Content Knowledge Test Performance by Students One way to assess improved learning is to conduct a content knowledge test of students by incorporating some target concepts in a set of questions. The questions are given both before and after the activities are carried out. Although learning capability widely varies from one individual to another, such a test can demonstrate the general effectiveness of the given "hands on" activities. The instructor can use the result as a bench mark and then continuously revise the activities until a high improvement in learning is achieved. A set of questions were given to Physical Geology students at UNI to assess their overall learning. The test consisted of seven (7) questions, each involving a specific hydrologic concept. The test was given to two groups of students. One group fully participated in all activities relating to the monitoring well site and the other group did not take part in it. The results demonstrated an improved learning of those students who took part in the outdoor activities. The primary reason for this improvement was that each of the given hydrologic concepts was better presented to the students by using the laboratory as well as the field activities. The students were not only able to see what was being taught, but they were also able to ask effective questions and then look for an answer themselves. In every step of the learning process, they experienced how science worked.

The teaching methods applied in this project helped students develop scientific thinking skills early in their academic pursuit. Besides giving the tests, the new knowledge they acquired in hydrologie concepts was assessed by taking them to actual contaminated sites and assigning practical problems to solve. As an example, they were taken to anoxic surface water bodies and asked questions as to why the water body became anoxic and what could be done to restore them. They derived answers from the information that were provided to them on the area hydrology along with the new concepts they developed through the project activities. There were group meetings onsite, including debates and question/answer sessions. Their ultimate goal of the day was to develop learning models that could be assessed and validated by fellow students. They soon began to realise that science was not for "scientists" only, out it was for everyday people. In effect, this whole effort better prepared the students for more advanced classes.

Removal of Misconception : During a class discussion at UNI, a physical geology student mentioned that he was worried about an accidental gasoline spill near his private drinking water well. A few others commented that they were aware of similar problems in other areas. Initially, it gave an impression that the students knew that there was a link between a spill and the quality of water in a nearby well. As the discussions continued, it became clear that some of these students had a misconception. They thought that groundwater was like rivers of water in the subsurface, with rock layers above and below. It seemed difficult for them to understand that water moves through "pore spaces" of rocks. At the end of the semester, they all said that their ideas about aquifers had changed. They added that it wouldn't be possible to grasp the concept if they didn't do the porosity experiments in the lab and then actually get involved in the dye tracing experiments at the on-campus monitoring well site. Through these hands-on activities they had a better grasp of contaminant movement in porous media and their possible remediation scenarios. The "spill" example in effect dealt with changing preconceptions.

As the materials were presented, the students first compared it to the preset ideas in their mind and then applied their own critical thinking to reset the concept in their knowledge pool. It is not always an easy task for the students to change preconceptions. So, the instructors should welcome questions from students and encourage them to challenge the new ideas. If the new experiences don't fit their original concept, the curricular process must reconsider what had been assumed and then attempt to rebuild the conceptual structures.

Consequences of Groundwater Pumping

The following statistics are pretty familiar, yet immensely sobering. We know that farmers use most of our water, about two-thirds in the United States, but consider that total groundwater usage exceeded 40 trillion gallons in the year 2000. Groundwater now constitutes more than 25% of the nation's supply, and over haft of us in the United States drink groundwater. Since 2000, the United States has experienced unbelievable growth and sustained drought. Farmers, cities, homeowners, and mines have searched for new supplies, and have almost always settled on groundwater. Though staggering, these are the best statistics available from the United States Geological Survey, and they are woefully out of date.

I next want to offer an overview of water law. There is a profound disconnect between what the legal system permits and what the science of hydrology teaches. In the American East, the use of surface water is governed by the doctrine of riparian water rights. If you own a piece of property on a lake or a river, you have water rights to that lake or river. They are shared water rights correlative rights because you and your fellow riparian property owners share the common resource. In the American West, settlers developed a different rule the prior appropriation doctrine. The motivation for that doctrine is sometimes attributed to the aridity of the West, but other systems used different rules. The Church of Jesus Christ of Latter Day Saints (the Mormons), Native American communities, and acequias in Northern New Mexico did it differently. The European settlers who moved west thought of water like other natural resources as something to be used instrumentally, productively.

The prior appropriation doctrine began in California in the 1850s, during the gold rush days. Miners found out that it took a lot of water to mine gold. They realised they needed to develop a system of rules to figure out which miner had rights to what water. There's a wonderful irony in this, because, if you think about it, these miners were thieves. They were stealing gold. Whether they were stealing from the federal government, the native peoples, or Spanish-speaking people is a fair topic to debate, but it definitely wasn't the miners' gold. Yet, the first thing these thieves did was to set up a legal system. So, the prior appropriation doctrine (first-in-time, first-in-right) awards rights to use a specific quantity of water with a specific diversion point, date, and purpose. That's the prior appropriation system.

Now comes the disconnect. When we shift from surface water to groundwater, the rules change, though not in every state. Oregon and a few other states have a version of the prior appropriation doctrine for groundwater as well as surface water, which integrates the two systems. Other states govern groundwater with a right of capture. Like the first case you read in property, Pierson v. Post, (1) involving wild animals, the right of capture considers groundwater a wild resource. If you can get it out of the ground, it's yours. Most states have a third rule, the reasonable use doctrine, which sounds good, but is an oxymoron. It allows a person to pump a limitless quantity of water as long as it is for a beneficial use. However, anything can be a beneficial use. The right of capture and the reasonable use doctrine epitomize the tragedy of the commons. Consider an aquifer as a giant milkshake glass. What the right of capture and the reasonable use doctrine permits is a limitless number of straws in the single glass. This is an absolutely maniacal way to run a system, and profound consequences can result from this. Eventually you run out of water; the finite quantity in the glass will be exhausted. Because the rules give everyone an incentive to put a straw in the glass, the supply is threatened.

That is what has happened with the Ogallala Aquifer. It stretches from the Dakotas in the north all the way down to the Texas Panhandle in the south. It had abundant groundwater resources, so much so that the area became the bread basket of the United States. However, the groundwater table has plummeted to such an extent that some farms have had to go back to dry land farming, creating a great deal of concern about the economic viability of the region. That's what happens when you allow limitless access to a common pool of resources; you will eventually run out. Even if you don't run out, you've got other problems along the way, one of which is increased energy costs. Those who hike know that every time you put that one quart Nalgene bottle in your bag, backpack, or daypack, you're adding two pounds of liquid. In the West we don't talk about water in terms of quarts or gallons. We talk about an acre-foot of water, which is the amount of water that it takes to cover an acre of land to the depth of 12 inches. That's 325,000 gallons. 325,000 gallons weighs 1,358 tons. If you're a farmer and you're pumping from 500 feet below the surface of the ground, you're looking at multi-thousand dollar electric bills for each well, each month. That's more than enough to drive farmers out of business.

As the water table declines, so does the water quality. This seems to be a function of the Earth's internal temperature rising as you pump from deeper levels where the water is warmer. Warmer water allows some nasty, naturally occurring chemicals to dissolve, such as fluoride, arsenic, and radon. Then there's the problem of salt water intrusion along all coastal areas. There is no place around, except perhaps the Oregon coast, that does not deal with this problem. Groundwater pumping causes salt water to migrate laterally and contaminate the potable supply.

Then there's the problem of subsidence. The surface level in parts of California and Arizona has dropped since 1925. How has that happened? Remember the last time you went to the supermarket and bought a box of Kellogg's Corn Flakes? You brought it home, opened it up, and your first thought was, "Kellogg's ripped me off." Before you had eaten one bowl of cereal, it was a third gone. If Kellogg's had been kind enough to have added two cups of milk to the box of cornflakes, they'd be way up to the top of the box. But if you take out the milk, the cornflakes settle. That's subsidence: remove the water and the land settles.

Other people have written about subsidence and groundwater intrusion. Water Follies is the first book ever published to focus on the environmental consequences of groundwater pumping: impacts on our surface waters, our lakes, our rivers, our creeks, our wetlands, and our estuaries. Let's start with Tucson, Arizona. As you drive west on Speedway Boulevard, there is a sign that reads: "Santa Cruz River." Tourists look over and start laughing. It's absolutely dry. It's not a river; it's a dry sand bar. Tourists think Tucson residents have been in the sun too long. But the sign makes sense if you look at it in historic terms. In 1942, there was a river at this location, flowing water and an immense cottonwood and willow gallery forest. But by 1989, nothing was left. All the trees were gone; all the water was gone; all these changes are a result of groundwater pumping.

How did this happen? The basic science of hydrology can be summed up with one principle: water only moves by the force of energy. The energy can come from the sun, evaporating water off of oceans or lakes. The energy can come from wind, as the water moves across the sky. The energy can come from gravity as water precipitates in the form of rain or snow. The energy can come from gravity as water flows down over the surface of the ground. It can come from gravity as water infiltrates the ground. It also comes from gravity as the water makes its way downhill to provide flow to the river. If it hasn't rained recently, water makes its way into a river from the ground. The water then flows laterally, subsurface to provide flows to our rivers and streams. A healthy river with a water table above the river stage supports the river with groundwater. But if the water table adjacent to the stream is lower, the water flows in the opposite direction. This is gravity nothing more, nothing less. Water seeks the lowest ground and flows to the lowest point. When we drill groundwater wells, we add another phenomenon. Groundwater wells intercept water that is on the way to the water course, but because of the pumping, will never arrive at the water course.

Another Arizona example involves the San Pedro River in southeastern Arizona. The San Pedro River is one of the fabulous birding places in North America. It's the meeting point of the southernmost stopping point for many species of Canadian and northern U.S. birds and the northernmost stopping point for many birds from South and Central America. It's an extraordinary place and the birds come due to the San Pedro River. Growth is out of control near the San Pedro. Whether it Hill go the way of the Santa Cruz River in Tucson is unknown, but in August 2005, the San Pedro went dry for the first time in recorded history. My message today is not just about Arizona, arid lands, or the American West. The environmental consequences of groundwater pumping are a national, and indeed, an international problem. My book happens to be about the United States. Let's first turn to the Midwest. I would like to share with you some stories of water follies. The first one occurs in the state of Wisconsin. Wisconsin is a state with great natural resources. Minnesota brags about having 10,000 lakes. Wisconsin has 15,000 lakes and seven to eight thousand miles of rivers. They have an active citizenry the tradition of La Follette Progressivism is alive and well in Wisconsin reflected in a major campaign to protect the Mecan River.

In the late Nineteenth century, the State of Wisconsin set out to protect the Mecan River. The state bought up huge sections of the river, rehabilitated portions, and obtained conservation easements from neighbouring landowners. As a consequence, the river is a blue-ribbon trout stream with naturally reproducing strains of brook, brown, and rainbow trout. Yet it faced a threat from a most unlikely source: bottled water. Think about the concept of bottled water. Where did it come from? How did it overwhelm us? No one used to drink bottled water in the United States, except perhaps at the office cooler, where you shared gossip or a dirty joke, or at Italian restaurants, or maybe in Berkeley. Now everyone drinks bottled water. This consumer craze has taken over. In my classroom, next to every laptop is a bottle of water. You go to the gym and everyone on the Stairmaster has a bottle of water. In fact, Stairmasters are now made with a place for you to put the bottle of water. Go to movie theatres and bottled water is more expensive than the popcorn! Bottled water now sells for more money than milk, oil, gasoline, or things made with water, like Coke.

Humans have a limitless capacity to deny reality. So what's bottled water got to do with the Mecan River? It has to do with Perrier, a French company that sells green bottles of spring water. Perrier's U.S. company, now called Nestle Waters of North America, is a subsidiary of Nestle, the largest food manufacturer in the world, and Nestle Waters of North America is the largest bottler of water in the United States. Nestle may not be a familiar brand of bottled water to you, but the company sells Arrowhead, Calistoga, Ice Mountain, Poland Spring, Zephyrhill and Osarka. They have fourteen different brands and a 32% market share of the bottled water industry in the United States. The company has decided for marketing reasons that American consumers will find greater cachet, and therefore pay more, for water that is labelled "spring water" rather than "artesian water," "ground water," filtered water," or any of the other FDA-approved labels. To sell water as "spring water," you must locate the well that pumps the water next to the spring.

That brings us back to Wisconsin. Nestle wanted to drill a well next to a spring that flows into the Mecan River. The state of Wisconsin has the reasonable use doctrine. There was nothing the state could do to prevent Nestle from drilling a well. Nestle's well was going to pump between five and six hundred gallons per minute, every minute, of every hour, of every dảy in the year. That's 272 million gallons of water a year that they put in bottles and take away. The well was going to be located sixty feet away from the spring. The spring carried a flow of between three and five cubic feet per second, a spring you can easily step over. The question was, "What's this going to do to the spring?" The answer, according to the Perrier hydrologist, was nothing.

I thought this was kind of a surprising conclusion. I went back to the University of Arizona and talked with Tom Maddock, a hydrologist. I said, "The Nestle guy says this isn't going to impact the spring." Tom replied, "He's not a hydrologist, he's a hydrostitute." There are some honest debates about what the hydrologic model shows, about parameters, calibrations, and other factors. There are arguments that don't meet what lawyers call the "straight face" test: if you can't make an argument on behalf of your client without smiling, you probably ought to forget that argument. Nestle's position doesn't meet the straight face test. It was absolutely absurd. The pumping would have devastated the spring. The reason why it's a water folly is that if Perrier had moved its well two miles away from the spring, the well would have produced water with the same chemical composition and had negligible impact on the spring. However, the water wouldn't have fit the company's marketing strategy of selling spring water. American consumers' fetish for spring water drove a company to such extremes.

In the end, this particular deal didn't go through because the aggressive citizenry of Wisconsin said it was unacceptable. The uproar from the local community was finally enough to convince Nestle to go next door to Michigan and open a bottling plant there. That plant has now been challenged, as have proposed plants in Maine and California. I want to turn now to Florida. Though Florida was once a swamp, those wetlands were drained long ago and they now have a water shortage problem. This is most improbable, because the state gets more than fifty inches of rain a year. On the west coast are the cities of Tampa and St. Petersburg: both big cities, both on peninsulas, both built out, both desperate for new water sources. In the 1970s and '80s, they went north and bought up huge tracts of rural land in Pasco County. Then they began to pump, and pump, and pump some more. Eventually they pumped so much that they created major problems, such as subsidence. This isn't Kellogg's Corn Flakes settling, this is Karst limestone. If you move water through limestone, you literally dissolve the structure. Scores of lakes in Florida, such as Crooked Lake, have dried up from groundwater pumping.

Consider the story of Steve and Kathy Monsees, Midwestern kids who were living in Florida. He was in the military and served in the first Gulf War. They liked Florida and decided to retire there. They bought a piece of property on a lake and built their retirement home. After his last posting from the Sudan in the early 1990s, they began living in their house on the lake. However, the lake became smaller and smaller until it finally dried up. Their retirement home on the lake had become mud flat property. At that point, they did what you are supposed to do in a democracy. They attended meetings, wrote letters, and tried to become informed about this issue. The utility just stonewalled them. Eventually it became obvious that these big well fields in Pasco County, right down the street from where their house was, had pumped all the water out from underneath them. At that point, the utility had a major public relations disaster. It had to do something. So the utility decided to refill the lake using groundwater. They pumped 375,000 gallons of water each day and dumped it into the lake, where it promptly percolated into the ground. It was like trying to keep water in a colander: madness.

The utility's engineers refused to recognise their folly, instead insisting their plan simply needed tweaking. They proposed to line the lake with an impermeable material and then dump the water in it. This is one of my favourite follies: engineers totally, blissfully ignorant. When the Water Follies book came out, Island Press received a twelve-page letter from the chief engineer, demanding that the book be taken off the market and defending the practice of refilling lakes as perfectly sound. Some people are clueless.

Let's move on to Texas. This story is about San Antonio, with its well-known River Walk. It's a vibrant downtown. The San Antonio River flows through, and boats go up and down the canals. There are fancy hotels, restaurants, and boutiques. Mariachi bands entertain the tourists. It's become the greatest tourist attraction in Texas, outdrawing even the Alamo. There's only one problem. There is no river. The river you see is a complete illusion. It's the dry San Antonio River bed, into which the city dumps up to 10 million gallons a day of groundwater pumped from the Edwards Aquifer in order to create the economically useful fiction of a river. The city circulates the water around, supplements it as needed, and, once a year, drains it and removes the silverware and the beer bottles because the "tourists like it natural."

The City of San Antonio is the largest city in the United States that is 100% dependent on groundwater. They are desperate for supplies. This led to the realisation that if the water in River Walk is such good water, why uses it only once? So the city now serves it to customers, runs it through their waste water treatment plant, then dumps it into the river. Tourists don't seem to have noticed a change. It's still groundwater, but it's been used once. San Antonio has got to find new water supplies.

Until recently, Texas had the right of capture doctrine. Under this doctrine, why would I buy your water right when I can go right next to your place and put in a well and suck your water out? That's called "the American way." It's not very neighbourly, but that's the right of capture. If you think about it, it's the very antithesis of a property right. It's akin to a circular firing squad, everyone's gun pointed in a mass, self-destructive act.

Under the hammer of the Endangered Species Act, (3) because there are some endangered species in the springs that discharge from the Edwards Aquifer, a federal judge ordered the state to change things. The state created an authority to divide up water rights. This meant the possibility of doing some deals. So, San Antonio first approached a local kid who went into the water business. He teamed up with a slumlord from New Jersey to find water. The Edwards Aquifer is a very productive aquifer. The area gets rainstorms measured in feet. In some places, there's even artesian pressure. Imagine an oil well gusher going up in the air. That's artesian pressure. By golly, these guys hit it. They put in a well with a diametre of thirty inches, and it rumbled like a freight train and erupted. It sent a column of water forty-five feet straight up in the air with a diametre of thirty inches. This thing put out 43 million gallons of water a day. So the first thing these guys did as religious people who were thankful for their blessings was name the well Ave Maria #1. I could just see a whole rosary of Ave Marias: Ave Maria #17, Ave Maria #22. They were on to something really big. Then they had to figure out what to do with 43 million gallons of water a day. They decided to raise catfish. To put this in perspective, this is one-quarter of the needs of the Sixth largest city in the country, and these turkeys are growing catfish and polluting the local streams to boot. It was so over the top, even in Texas, that the city came in and bought them out for 10 million dollars. Water has a value we haven't appreciated.

The city still needed more water so they turned to the panhandle specifically Roberts County and to a fellow by the name of T. Boone Pickens, quite a legendary character in Texas. Pickens said to San Antonio, "Look, you need water; I've got water. I'll sell you the water from beneath my ranch. I will pump it out and sell it to you." However, there was little water beneath his ranch. Remember the Ogallala Aquifer? That's where his ranch is. What he was really volunteering to do was to sell the water out from underneath his neighbours' ranches, because that's what the right of capture permits. Your pumping creates a gradient, and you just sort of drain this water from neighbouring property, and sell it to San Antonio.

When we in the American West think about the Central Arizona Project, about moving water hundreds of miles and projects costing billions of dollars, we think blissfully about something else. That something else is the United States Treasury. We think about taxpayers in New Jersey who haven't a clue about western water projects. In Arizona, we think that the role of the federal government is to bring in wheelbarrows full of money, drop them, and then go away. We are independent, free spirits in Arizona. This project, however, didn't have any federal money, didn't have any state money, didn't have any local money. It had no government support whatsoever. Pickens was willing to put up a billion dollars of his own money to do this deal. Water has a value we haven't appreciated.

Let's turn to Minnesota, a great state with wonderful natural resources. This story is about McDonald's french fries. We've all eaten McDonald's french fries; the average American consumes thirty pounds of fries a year. This story is about Ray Kroc, the founder of McDonald's. He was a marketing genius who understood the whole idea of fries. If you want to make a great fry, you've got to start with a potato that has a high water content, because the process of making the fries is the process of extracting the water and replacing it with fat. That's why they taste so good: they contain lots of fat and taste good, especially with a lot of salt. So how does this relate to Minnesota?

It relates to Minnesota in that a sea change in water use in the United States is occurring as farmers in Minnesota are irrigating. Traditional dry-land farmers in the Midwest, the East, and the Southeast are now irrigating to get a higher yield from their crops. This story revolves around Ron Offutt, one of the largest potato growers in the United States, who also happens to be the largest John Deere dealer in the United States. What does potato farming have to do with the environment? It has to do with the Straight River.

The Straight River is a blue-ribbon trout stream threatened by using water to grow potatoes. Potatoes grow perfectly well elsewhere without the need for irrigation. Farmers have been growing potatoes in Maine for hundreds of years, and you can make fries with potatoes grown this way. But when you make fries without irrigation the potatoes could be slightly different, not the same. The next time you go to the supermarket, try to find real potatoes. They are mostly found in the frozen food section. You can find thirteen types of frozen french fries more models than GM has. Every potato is the same as the next one: same diametre, same colour, and same length, which for McDonald's purposes, is most important. In the 1980s, McDonald's went from selling french fries in little waxed paper bags, to selling them in the hard-sided boxes that we now know well Super Sized. From that point on, McDonald's would only buy from farmers who irrigated, because irrigation produces what are called in the trade, "industrial potatoes." They have the same uniformity as a wristwatch or a computer, and are just the right length to stick out of that Super Sized carton and be grabbed between your thumb and forefinger and dipped in catsup. That's a water folly. Potatoes can grow perfectly well without irrigation. In fact, they do so for markets in Japan, and consumers in Japan are notoriously fastidious about their foods. Yet, in America all our fries are the same length, perfectly white and blemish-free. As a final story, let's return to Arizona and the Grand Canyon, a fabulous place. But we are loving the Canyon to death. Every year five to six million of us visit the Grand Canyon, and at this point the place has problems. There's haze over the Canyon from the air pollution from mining operations, traffic congestion is a problem, and the housing facilities for park employees and concessionaires are in appalling shape. We have to do better by this crown jewel. We simply must.

This story is about the future of water. It begins with a Scottsdale, Arizona developer who wants to do a deal. In the little town of Tusayan, just outside Grand Canyon National Park, he wants to do a land swap with the Forest Service. He owns 2,200 acres of land inside the National Forest, and he wants to exchange that land for 262 acres of land in the town of Tusayan owned by the Forest Service. The proposal was for a mixed-use development with some condos, some motel units and some commercial sites, a Native American craft centre, and housing for park employees. He proposed to supply the project with groundwater. To do this swap required preparation of an environmental impact statement (EIS). The EIS revealed that using groundwater would reduce the flow in Havasu Spring and in Indian Garden Creek.

Indian Garden Creek is a perennial creek which supports an amazing cottonwood oasis that you encounter halfway between the

South Canyon rim and the river. It's a place to stop, fill up your canteen, relax in the shade, and enjoy the splendor of the Canyon. Those who have rafted the Canyon know that Havasu Spring comes in two-thirds of the way down the Canyon, river left. It's a turquoise river with turquoise waterfalls. It's an extraordinary place. It looks like it should be in the Caribbean, not in Arizona. When the EIS was released, indicating that Indian Garden Creek and Havasu Spring would be affected by the proposed groundwater pumping, it generated an angry reaction. How could you think of doing something that would have that kind of impact on the Grand Canyon! The developer knew this was not an idea that merely needed tweaking. It was back to the drawing table. The developer pulled together the major stakeholders: Grand Canyon National Park, the Forest Service, the two principal tribes, the Grand Canyon Trust, and the National Parks and Conservation Association. He brought these players to the table to figure out a different water supply. They proposed having the developer buy some Colorado River water rights, which he would divert at Topock. Topock is further south, on the western border of Arizona. He would load water onto Burlington Northern tankers, and move the water 180 miles by railroad to Williams, where he would offload the water. At that point, he would put the water in a pipeline that he would build for 15 million dollars and move it fifty miles from Williams to Tusayan, where he would use the water.

This was going to cost a lot of money. After adding up all the numbers, it came to 20,000 dollars per acre-foot of water. Farmers in the Welton-Mohawk Irrigation District in southwest Arizona, pay 15 dollars an acre-foot; across the border in California, farmers pay 15 dollars in the Imperial Irrigation District for that same Colorado River water. This was 20,000 dollar water. Farmers can't make a profit if they have to pay 20,000 dollars, even if they grew marijuana. The developer, bewildered by these numbers, gave them to his accountant who crunched them and said: "It's a go."

Let's stop thinking about water in terms of acre-feet and start thinking about water in terms of Nestle. 20,000 dollar water ends up being six cents a gallon. It's just a cost of doing business for a development at the gateway to one of the grand splendors on earth. Water has a value we haven't appreciated.

The cause of these problems: population growth. That's the basis of every environmental problem. Every hour, the population of the state of California increases by about sixty people. The population of the United States is now 300 million, up 15 million in the last five years. Demographers predict we will hit 400 million by 2043. It's all about population. The problem also involves a tragedy of the commons, brought about by the profound disconnect between principles of hydrology and legal rules. Hydrologists understand the problem: the laws are out of whack. We've got to break that relentless cycle. It is absolutely bewildering to think that we let anyone who wants, under the guise of property rights, just put more straws into the milkshake glass. These stories are funny but also quite serious. The impact of groundwater pumping on the environment is hidden, and it occurs slowly over years or even decades. Groundwater pumping that has already occurred is going to have consequences on the nation's fresh waters in years to come. We must take action. The time is now. We need to recognise that water is both a public resource and has aspects of private property. We can and should encourage water conservation. Some places, such as Tucson, have already gone quite a ways down the path of conservation, but there are limits to what we can achieve through conservation. What has not been done with water policy in the United States is to consider market- based solutions, price signals, and incentives. We need to tell every new user of water, if you want to put a straw in the glass and make the problem worse, then you've got to take some other straw out. You've got to buy a current user's right to use water. We will no longer permit you to make the situation worse. This market process will help to move water from lower value to higher value uses.

We must also begin to price water appropriately. Most Americans pay more money for their cell phones and cable TV than for water. We need to raise the price of water. This is a very dicey political issue, but we need to begin a dialogue about how to structure water rates appropriately. Let's begin with a lifeline rate. In the richest country in the history of the world, we can surely recognise a human right to water. But that amount of water, calculated at twelve to fifteen gallons per capita per day, only constitutes 1% of the total water use in the United States. That leaves the other 99% to be accounted for. That's where market forces can play a role, with voluntary transfers between willing sellers and willing buyers. There is reason to be optimistic. Mother Nature is remarkably forgiving and regenerative, although we have treated her resources shamefully. If we can slightly redirect our water policy, there is reason to think that the springs will bubble and the rivers will flow.

Bibliography

Anderson, Mary P. and William W. Woessner: *Applied Groundwater Modeling: Simulation of Flow and Advective Transport.* Academic Press, San Diego, 1992.

Asit K. Biswas and Cecilia Tortajada: *Appraising Sustainable Development: Water Management and Environmental Challenges*, Oxford University Press, Delhi, 2005.

Bhakar S.R. *: Ground Water Hydrology : Theory and Practice*, Agrotech, Delhi, 2009.

Bhave, P.R. and R. Gupta: *Analysis of Water Distribution Networks*, Narosa, Delhi, 2011.

Boyer , J.S.: *Measuring the Water Status of Plants and Soils,* Academic Press, N.Y., 1995.

Conway, D.: *Climate Change and Water Resources in the Nile Basin*, University of London, London, 1993.

Cullis, A. : *Rainwater Harvesting: The Collection of Rainfall and Runoff in Rural Areas,* London, U.K.: IT Publications, 1986.

Doneen L.D.: *Water Quality for Agriculture*, Department of Irrigation, University of California, Davis, 1964.

Engelman, R., and P. LeRoy : *Sustaining Water Washington*, D.C.: Population Services International, New York, 1993.

Freeze, R. Allan and John A. Cherry: *Groundwater*. Prentice Hall, Englewood Cliffs, NJ. 1979.

Ghosh, N.C. and K.D. Sharma: *Groundwater Modelling and Management*, Capital Pub, Delhi, 2006.

Gupta K.R.: *Water Crisis in India*, Atlantic Pub, Delhi, 2008.

Heath, Ralph C.: *Basic Ground-water Hydrology*, Denver, CO: U.S. Geological Survey, 1983.

Husain, Ahmad : *Environment and Water Resource Management*, Sumit Enterprises, Delhi, 2006.

Jana B L : *Water Harvesting and Watershed Management*, Agrotech, Delhi, 2008.

Kally, Elisha and Gideon Fishelson: *Water and Peace: Water Resources and the Arab-Israeli Peace Process*, Westport, CT, Praeger, 1993.

Linda M. Welkom: *Groundwater Chemicals Desk Reference*, Chelsea, MI, 1990.

Lowi, Miriam: *Water and Power: the Politics of a Scarce Resource in the Jordan River Basin*, Cambridge University Press, Cambridge, Eng, 1993.

Matthess, Georg : *The Properties of Groundwater*, John Wiley & Sons, New York, NY. 1982.

Meenu Bhatnagar: *Groundwater Management : Sustainable Approaches*, Icfai Books, Delhi, 2012.

Montgomery, John H. and Linda M. Welkom: *Groundwater Chemicals Desk Reference*, Chelsea, MI, 1990.

Nielsen, David M.: *Practical Handbook of Groundwater Monitoring*. Lewis Publishers, Chelsea, 1991.

Pacey A., and A. Cullis : *Rainwater Harvesting: The Collection of Rainfall and Runoff in Rural Areas,* London, U.K.: IT Publications, 1986.

Parikshit Ballabh: *Water Resource Management : Planning and Development*, Cyber Tech, Delhi, 2009.

Rao K. Nageswara: *Water Resources Management : Realities and Challenges*, New Century Publication, Delhi, 2006.

Rowe, Garry W. and Sylvia J. Dulaney: *Building and using a Groundwater Database*. Lewis Publishers, Chelsea, 1991.

Sharma, S.S.P. and U.H. Kumar: *Dynamics of Watershed Development and Livelihood in India*, Serials Pub, Delhi, 2010.

Singh A K : *Environment and Water Resources Management*, Adhyayan, Delhi, 2006.

Thapliyal, B K : *Democratisation of Water*, Serials Pub, Delhi, 2008.

Tideman, E. M.: *Watershed Management – Guidelines for Indian Conditions*, Omega Scientific Publishers, Delhi, 1996.

Walton, William C.: *Groundwater Resource Evaluation*. McGraw Hill, New York, 1970.

Index

I

L

M

N

P

Q

R

S

U

W

❑❑❑